Mars: The Living Planet

MARS
The Living Planet

Barry E. DiGregorio

with

Dr. Gilbert V. Levin and
Dr. Patricia Ann Straat

Frog, Ltd.
Berkeley, California

Published by Frog, Ltd.

Frog, Ltd. books are distributed by
North Atlantic Books
P.O. Box 12327
Berkeley, California 94712

Cover photos:
  Micrograph of snow algae by Dr. Ron Hoham
  Inset photo of Mars courtesy of the Hubble Space Telescope Science Institute
Cover design: Paula Morrison
Interior book design: Catherine E. Campaigne

Printed in the United States of America by Printer

Library of Congress Cataloging-in-Publication Data

DiGregorio, Barry E.
    Mars : the living planet / Barry E. DiGregorio.
       p.   cm.
    ISBN 1-883319-58-7
    1. Mars (Planet)   2. Exobiology.   3. Life on other planets. I. Title.
    QB641.D54   1997
    576.8'39'099923—dc21   96-37580
                                                                    CIP

1   2   3   4   5   6   7   8   9   /   02   01   00   99   98   97

To Gilbert V. Levin and Patricia Ann Straat, discoverer and co-discoverer of life on Mars in 1976, and to William M. Sinton, who led the way.

Table of Contents

Foreword

I had the great honor of working on *Viking Lander* data when I was invited as a student by the Jet Propulsion Laboratory. (I did not know then I was to be the very last student working inside the Viking team in JPL, as this fantastic mission was close to its end!) At that time, I was much more interested in looking at the wonderful landscape images returned to Earth each week by Lander 1 than trying to understand the question of Martian life. And who cared any more about Martians? NASA said there was no life on Mars....

Working my way through piles of scientific documentation, I discovered, however, that among the three Viking biology experiments, one (Dr. Gilbert V. Levin's Labeled Release experiment or "LR") did give strong positive results on Mars. This was a shock for me, because suddenly the horizon was widening on my desert images.... Dr. Levin's LR was the most sensitive of the three experiments done on Mars. It has been successfully counter-tested with Antarctic soil and Mojave Desert among others and could detect as few as 10-50 bacterial cells per gram... In contrast to the LR's sensitivity, the Gas-Chromatograph Mass Spectrometer (GCMS) mounted inside the *Viking Landers* would need 10,000,000 bacterial cells to register a response and therefore would have been unable to detect organic compounds in a desert soil containing fewer numbers. As a result of those counter-tests, the LR found life on Earth, while the other two experiments and the GCMS did not!

Maybe small dormant Martian bacteria have been mixed with soil by erosional agents for aeons and are now covering large parts of the planet (one theory), remaining unchanged for millions of years until some of them were suddenly revived inside the two LR instruments on board *Viking Lander 1* and *2*.

Dr. Levin also documented strange color changes in the soil and rocks at the *Viking Lander 1* site. One rock, called "Patch Rock," was of particular interest to him due to color changes that seemed to intensify during the spring-summer seasons. I am not sure about the nature of these changes: having processed a lot of Lander pictures, I saw that rocks can undergo changes in color due to deposition and removal of dust after dust storms. But that does not mean there is no life on Mars. As Gerald Soffen (Viking Project Scientist) told us in his book *The Viking Project*, "All the signs suggest that life exists on Mars, but we can't find any bodies." Obviously most scientists feel those microscopic bodies were not seen by the Landers' cameras.

Now I think that Dr. Levin and his co-experimenter Dr. Patricia Ann Straat, after all these years, may have discovered life on Mars after all, and could be recognized soon as the very first discoverers of life on another planet.

—Olivier de Goursac, Executive
Director, Planetary Section,
Société Astronomique de France
(SAF), Paris, France,
December 1, 1996

Preface

I have been fascinated with the Mars-life issue ever since 1964, when my mother and father gave me a book entitled *Mars* by Robert S. Richardson and Chesley Bonestell. I remember the book having the story of Dr. William M. Sinton and his infrared search for vegetation on Mars with seemingly positive results. Then the story slowly faded when Dr. Sinton retracted his announcement of life on Mars because of his scientific peers, who insisted that what he was observing was nothing more than water in the Earth's atmosphere.

By this time, I was an experienced amateur astronomer with several different sizes of telescopes ranging from a small 3-inch refractor to a giant 12½-inch reflector. The Moon and Mars were my specialty. I observed and photographed the Moon and Mars whenever they were available for viewing. I amassed a huge collection of photographs through the years. I also began to study microbiology.

Then along came the Viking Mission to Mars in 1976. My first real interest in the Viking biology results came from reading an article in *National Geographic* magazine. In its January 1977 issue was an article entitled "Mars: The Search For Life." The article painted an optimistic view that life on Mars might have been discovered. This was where I read about a biology experiment called "Gulliver" and later the "Labeled Release" (LR) experiment, and the scientist who conceived it, Dr. Gilbert V. Levin who, with the help of his coworkers, principally Dr. Patricia Ann Straat, moved the concept to reality.

On and off through the years, I would read an occasional article or interview with Dr. Levin. His LR experiment had gone to Mars and returned results indicating the presence of living microorganisms. However, his data were downplayed by NASA. He would insist that his LR data was never fully reviewed. It seemed that whenever I read an article by others on the Viking biology results from Mars, whether in prominent scientific journals or amateur astronomy magazines, their claims were all the same—that "no evidence of life was found by the Viking biology experiments."

This denial continued right up until the announcement by the National Aeronautics and Space Administration of possible fossilized bacteria found in the Martian meteorite ALH 84001 on August 7, 1996. The stories about life on Mars were slowly changing, and NASA now admits there could have been life on Mars 3.6 billion years ago but not today. Then this changed again as a team of British scientists analyzing a younger meteorite from Mars (EETA 79001) also found evidence of biologically produced materials, and suggested life on Mars may have existed as recently as 600,000 years ago and may have continued until today!

When I sought out Dr. Gilbert V. Levin and learned what he and his co-experimenter, Dr. Patricia Ann Straat, had to go through with NASA, and the utter humiliation they had to contend with from the NASA scientific community in spite of the fact that they had good data suggesting life on Mars, I was astonished that no one seemed to care. Here was possibly the greatest discovery in the history of science, but most scientists, scholars, and writers became Mars-life shy. Why? After Viking no one wanted to be associated with anyone who claimed there was life on Mars, for fear of being branded a Percival Lowell. Everyone wanted to play it safe and ride on the politically correct "dead Mars" bandwagon.

As I studied the scientific papers and information contained in Dr. Levin's files, a strange picture of NASA began to emerge which indicated to me that the agency did not want anyone to know whether life on Mars was discovered. I questioned why and decided to write this book, *Mars: The Living Planet*, before NASA foolishly attempts to bring back a soil sample from Mars and risk contaminating the Earth with possible pathogenic bacteria or viruses.

I hope I have rendered the story of Dr. Gilbert V. Levin and Patricia Ann Straat with the dignity and respect they deserve as scientists. They are, more likely than not, the discoverers of life on another world—Mars. I recall how the scientific geniuses of antiquity had to endure persecution from peers and clergy —great thinking people like Copernicus, Galileo, and the incredibly brave Giordano Bruno, for merely telling the truth about nature. It might not have been a truth that was convenient, nor one that everyone wanted to hear, but it was truth nevertheless. They believed, as all great scientists do, that every thinking person deserves a chance to experience new truths in nature whenever they are found. Giordano Bruno, while being burned alive at the stake by seventeenth-century clergy, cried out as his last words that the universe was filled with other solar systems which have planets, and that life must exist throughout the cosmos. How many of us would treasure truth and beauty until such a bitter end?

If this book seems slanted or biased in the direction of Dr. Levin and Dr. Straat, forgive me. It is only because of all those countless articles that have been biased or slanted in the direction of a chemical explanation for what happened on Mars in 1976 with Viking. With new evidence that life existed on Mars billions of years ago, there is no scientific reason why that life may not have continued to evolve right up until today. Yet, it is repeated year after year that "the Viking Lander Biology Experiments searched but did not find any evidence of life on Mars," as the quote reads that is emblazoned on the Viking exhibit at the Smithsonian Institution. That statement is both unscientific and untrue. The fact that this claim has been perpetuated by NASA for more than twenty years despite evidence to the contrary is nothing less than suspicious. There is only one way to be certain about life on Mars, and that is to return to Mars with updated versions of life-detection experiments designed to settle that issue once and for all. Since Viking, NASA has turned its back on settling this incredibly important issue. Why?

—Barry E. DiGregorio
Middleport, New York, 1997

Acknowledgments

Writing *Mars:The Living Planet* was a challenging project, but as with all substantive works, it involved other contributors. I have the following people to thank for their outstanding support, comments and photographs: First and foremost, I thank Dr. Gilbert V. Levin for providing the inspiration for me to write this book and allowing me access to his personal files and photographs dating from the Viking Mission to Mars. Dr. Levin also volunteered graciously to read the manuscript and contributed many useful comments and suggestions. I must also thank you, Dr. Levin, for your scientific and technical editing of the areas in the manuscript where it pertains to the Viking Mission and of course, for Chapter Nine: "Life After Viking: The Evidence Mounts," containing your special new announcement.

I wish to thank Dr. Patricia Ann Straat, who was co-investigator of the Viking Labeled Release experiment and is currently Deputy Chief for Referral, Division of Research Grants of the National Institutes of Health in Maryland, for her useful comments and suggestions pertaining to Dr. Wolf Vishniac's important work in the "Great White Desert," and her experiences working with Dr. Gilbert V. Levin and the Viking Biology Team. Your closing comments for *Mars: The Living Planet* are very much appreciated as well, Dr. Straat.

I was honored to get firsthand comments and insights from telescopic-infrared pioneer Dr. William M. Sinton of the Lowell Observatory, who in the 1950s and 1960s created a surge of interest in Mars with his announcement of "vegetation on Mars." I was able to talk with Dr. Sinton at length before his serious bout with Lou Gehrig's disease, which left him paralyzed from the neck down in 1996. Dr. Christopher P. McKay of the NASA Ames Research Center provided me with many valuable insights into the nature of life in extreme environments on the Earth (hot and cold deserts) where he has conducted much of his ground-breaking research. Dr. McKay provided additional information on Antarctica, hydrogen peroxide, and the superoxides so crucial to understanding life on Earth as well as on Mars. Thank you for the photographs as well, Dr. McKay.

Dr. I. Emre Friedmann of the Florida State University Polar Desert Research Facility also gave freely of his most significant research papers on microorganisms and harsh climates, including the work he has done with the 300-million-year-old bacteria found alive in Siberian permafrost. Drs. Friedmann and McKay have worked together to provide science with cutting-edge knowledge important to exobiology and Mars. I wish to thank Dr. Ronald W. Hoham of the Department of Biology at Colgate University in Hamilton, New York, for his fascinating comments and suggestions on the world of the snow algae. Dr. Hoham also provided the cover photograph (a micrograph he took himself) for *Mars: The Living Planet* showing various species of snow algae which may, some day, be shown to have relatives on Mars. I thank Dr. Klaus Biemann, the former Principal Investigator of the Viking GCMS experiment, for his kind comments and suggestion that *Mars: The Living Planet* carry a question mark at the end of the title (even though I am not using it for obvious reasons). Dr. Norman Horowitz, former Principal Investigator of the Viking Pyrolytic Release experiment, provided me with valuable comments and quotes.

Professor G.A. Avanesov of the Space Research Institute of the Russian Academy of Sciences in Moscow generously provided detailed information on the Russian MARS-96 Mission. A big thank you goes out to Rick Klein, Manager of the Cornell University SPIF facility, for his hospitality and allowing me access to all the *Mariner 9* and

Viking images there. My appreciation goes to Dr. Timothy Swindle of the University of Arizona Department of Lunar and Planetary Studies for providing valuable information on both the ALH 84001 and EETA 79001 Martian meteorites. Good luck on your "Life on Mars" class, Dr. Tim, it sounds very exciting.

I must thank my good friend Dorothy Schaumberg of the Mary Lea Shane Archives of the Lick Observatory for her perseverance in obtaining nineteenth-century historical information and drawings of Mars. Another professional who helped me obtain the latest color renditions of Mars from the U.S. Geological Survey Data Facility in Flagstaff, Arizona, was Adrianne Wasserman. Thanks also to Dr. Alfred McEwen of the U.S.G.S. at Flagstaff's Astrogeology Branch for comments on how new images of Mars are being reconstructed from archival Viking data.

Special thanks to Edwin Strickland III for his abstracts on the surface colors of Mars. It was Ed Strickland's ground-breaking work that, after Dr. Levin's pioneering work, demonstrated that Mars had many colors in the soil, even though Ed does not attribute them to biology. Dr. Terry Z. Martin of the NASA Jet Propulsion Laboratory provided thermal inertia data from his work on the Viking Infrared Thermal Mapper, showing that Mars has temperatures much in excess (at times) of those commonly published in magazines and journals. Dr. James Tillman of the University of Washington in Seattle, another of the Viking alumni who worked as Principal Investigator on the Viking Meteorology Team, provided much material on the temperatures and wind-speeds of Mars. Thanks to Dr. Thomas Duxbury of JPL for images relating to the Russian Phobos II Mission to Mars in 1988.

A big thank you goes out to Dr. Leonard Martin of the Lowell Observatory in Flagstaff, Arizona, for allowing me to include his fascinating photographic evidence of possible active geysers and other volcanic outgassing events on Mars. Another cutting-edge contributor to *Mars: The Living Planet* was Dr. Todd O. Stevens of Pacific Northwest Laboratories in Richland, Washington, who discovered the microorganisms (chemolithoautotrophs!) that live off a reaction between basalt rock and water deep within the aquifers of Earth. Dr.

Stevens provided valuable comments and interesting photographs of these organisms now thought to be one of the possibilities for current Mars life.

Many thanks to Dr. Paul W. Rundel of the Department of Biology at the University of California. Dr. Rundel has studied the ecology of the Atacama and Peruvian Deserts and discovered that entire ecosystems can survive on just the water vapor derived from fog. Mars has lots of ice-fog, Dr. Rundel. Thank you's are in order for Dr. James Bell of Cornell University for information concerning the possibility of organics on Mars, and the same thanks goes out to Dr. Yvonne Pendelton at NASA-Ames for all her comments and research papers on the infrared spectrum. Dr. Ted L. Roush of the Department of Geosciences at San Francisco State University was very helpful in providing diagrams that showed the evolution of research on Mars pertaining to the infrared search for organic material. Thanks to Dr. Helen Vishniac, Professor Emerita at Oklahoma State University's Department of Microbiology and Molecular Genetics, for your comments on not only your own work in Antarctica, but also comments provided on your late husband, the great Dr. Wolf Vishniac. Mary Noel Black from the Lunar and Planetary Institute obtained permission to use interesting images from the Center for Advanced Space Studies in Houston, Texas.

Special thanks to the American Geophysical Union who granted me permission to use any quotes, diagrams or photographs from their book *Scientific Results of the Viking Project* (1977), which contains reprints of the many scientific papers published (from the *Journal of Geophysical Research*) during the time *Viking* was sending its data Earthward from Mars.

I wish to thank William Sheehan, for his professional support and encouragement. Thanks is also in order to Olivier de Goursac, Executive Director of the Space Exploration Section of the Astronomical Society of France, for providing the Foreword to this book and images and text regarding work that he and Dr. James Garvin conducted at Brown University on Viking images showing changes on the surface as the seasons on Mars changed.

I would like to thank NASA and the people out at the Jet Propulsion Laboratory for various images and comments. Special thanks to Jurrie van der Woude of JPL for his detailed comments regarding his experiences in the Viking JPL control room when NASA administrator James Fletcher ordered the Mars "blue sky" image destroyed. A big thank you to my cousin, Gerry Clancy, from Burlington, Ontario, Canada, for creating a new computer program (in less than one hour!) that makes my system easily read by other computers, thus saving me weeks of additional work on the manuscript.

Thank you's are in order for my loyal family: my wife Susan and our two children Neil and Aurora, who did not see me often enough during the time it took me to write this book. I thank all the good people out at Color Tech Imaging in Amherst, New York, for their outstanding professional work and courtesy.

My only regret is that I will not be able to read Dr. Carl Sagan's review of my book in 1997. His sudden death in December of 1996 was a shock to all of us who believed his health was improving. Dr. Sagan, a science-hero of mine since childhood, had expressed enthusiasm about *Mars: The Living Planet* in a letter to me during the summer of 1996. While Dr. Sagan shared his interest with me in this project, he also voiced concern that my book would not "go beyond the data." In this respect, I assured him that my goal was to present as much data as possible regarding the Mars life issue, even if the data happened to contradict other scientific findings which have been popularized by the NASA scientific community. In a November 6, 1996, *Boston Globe* article entitled "Seeking Life Beyond Earth," Dr. Sagan called for a re-examination of the *Viking Lander* biology experiments, questioning whether the NASA-supported peroxide theories had ever explained any of the biology data. He also called Dr. Gilbert V. Levin "a lone voice in the wilderness" who has upheld through the years that his LR data from *Viking* suggested life on Mars. Sagan's last comments regarding Levin were—"he has a case."

Finally, there were many other individuals who in numerous small ways assisted, encouraged, and otherwise kept me going during the tough times. You know who you are, and to you, my sincerest gratitude.

Mars, Exobiology, and the Origin of Life: An Introduction

... [T]he microbial world has been on our planet for billions of years before we were here and will continue to be there for a very long time ... disease is an ever constant threat, it kills more people than do wars....

—DR. JOSHUA LEDERBERG (1996), PRESIDENT EMERITUS AND INFECTIOUS DISEASE RESEARCHER OF ROCKEFELLER UNIVERSITY

H ow, where, when, and why did life arise? These questions are perhaps the most important and elusive in the history of science, yet it has only been within the last fifty years that we have actually been able to look for direct evidence. Our tools are infrared spectroanalysis techniques pioneered by astronomers in the 1950s, new discoveries in microbiology, and the sending of robotic spacecraft to the Moon and planets.

In August 1960 the obscure new science of astrobiology was given a new name, "exobiology," by then-professor of genetics at Stanford University, Joshua Lederberg. Exobiology is the study of the origin, evolution, and distribution of life in the universe. Life is composed of matter, so the search for life begins with the early organization of matter.

Spectacular images returned by the Hubble Space Telescope in 1995 in a dusty star-forming region known as the Eagle Nebula almost seven thousand light years distant are veritable photographic proof that stars and solar systems are formed in gaseous clouds of molecular hydrogen. These regions have appropriately been designated "EGGs," an acronym for Evaporating Gaseous Globules. The surrounding molecular hydrogen gas from which each star is born is then "blown away" by the intense ultraviolet solar wind from the massive newborn star in a process called photoevaporation. As the new stars emerge, each may have an accumulation of dust swirling and condensing into a collection of baseball- to Moon-sized objects called planetesimals. These planetesimals grow into planets, and some become asteroids and comets. This is how a solar system is born.

But what of these nagging questions: From where does life arise? What period in solar system history does life initially get started? Does life begin in comets or on asteroids first?

The standard model for the origin of life on Earth, accepted as dogma by many biologists today, was first hypothesized by the Russian biochemist A. P. Oparin in 1924 and modified by a Scottish counterpart, J. B. S. Haldane, in 1928. According to this hypothesis the Earth had a primitive reducing atmosphere consisting of hydrogen-rich molecules from the leftover gases of the solar nebula that became trapped in our planet's gravitational field. This reducing atmosphere consisted of hydrogen, methane, ammonia, and water vapor. From these chemicals —along with electrical discharges in the form of lightning and intense ultraviolet light from the Sun—an abundance of organic chemicals built up in the primitive oceans of Earth and became a "soup" of energy-rich biochemicals. Interactions among these chemicals produced more and more complex organic molecules which "somehow" became self-replicating entities, or life. These early life forms would have no need to synthesize products of elementary metabolism like amino acids and nucleotides because all the nutrient requirements were provided by the oceans. It was only after all the nutrients in the oceans were depleted that life would need to evolve other means of subsistence.

According to Oparin, the first organisms that metabolized were the anaerobic organisms, early bacteria that could live only in an oxygen-free environment. These early organisms then learned to live off carbon dioxide in the ancient secondary atmosphere of Earth as their only source of carbon. Oxidizing sulfur and iron compounds as a source of energy, these organisms are called "autotrophic" because of their ability to grow without organic compounds in a minimal medium.

Eventually, some anaerobic microorganisms evolved to use the energy of the Sun and became phototrophs, organisms capable of photosynthesis. The earliest fossil evidence indicates that photosynthetic organisms were present 3.8 billion years ago. These cyanobacteria or blue-green algae would eventually convert the carbon dioxide atmosphere of our planet into the oxygen-abundant world we live in today. The newly oxygenated atmosphere of Earth wreaked havoc with microorganisms that lived on carbon dioxide as their only source of carbon, and most of these early life forms were exterminated by this new oxygen envelope that surrounded Earth. Then life took a new course as those organisms that could adapt to their new environment evolved to use it to their advantage. This "oxygen holocaust," as it has been termed, was a turning point in the evolution of life on Earth and demonstrated life's ability to change in accordance with a new and hostile environment. This is perhaps the most persuasive argument for the existence of extant life on Mars today, even though many scientists still believe life can no longer exist on Mars because of its thin atmosphere of carbon dioxide, limited water, cold conditions along with intense ultraviolet light reaching its surface, and a host of oxidizing chemicals which are postulated but for which we have no scientific evidence.

While the Oparin-Haldane hypothesis remained an intellectual piece of the evolutionary jigsaw puzzle for a quarter of a century, its fundamental characteristics were essentially proven in a 1952 experiment considered to be one of the most important in the history of science. Dr. Harold C. Urey, a physical chemist from the University of Chicago, and Stanley Miller, his graduate student, set out to test the Oparin-Haldane theory by taking it to its very basic level, that of the reducing atmosphere, which they simulated in a University of Chicago laboratory.

Their scientific apparatus was nothing more than a few simple pieces of chemistry hardware such as a five-liter flask with two tungsten electrodes inside. This was connected to a condenser rod that had a tube running into a 500-cc flask which had a stopcock permitting samples to be withdrawn while conducting the experiment.

The first phase introduced small amounts of water and mixtures of gases that included methane, ammonia, water vapor, and hydrogen without any oxygen, simulating the gases in the early atmosphere according to the Oparin-Haldane theory. The scientists then introduced repeated electrical discharges through this simulated primitive atmosphere and after only a week began to notice dark sedimentary material gathering in the lower flask. When Drs. Urey and Miller analyzed these sediments they were found to contain the rich organic molecules that Oparin-Haldane predicted would form. They found aldehydes, carboxylic acids, and the largest surprise of all, amino acids—in other words, the very basic molecules contained in all living cells on Earth.

New discoveries in planetary atmospheric physics have led astronomers and exobiologists to reconsider many aspects of the Oparin-Haldane theory, or as it is more commonly described among biologists today, "the Oparin-Ocean Scenario," and the spontaneous chemical origin of life on Earth. For example, the Oparin-Haldane theory depends on a primitive reducing atmosphere composed of hydrogen, methane, ammonia, and water vapor like those in the early solar nebula and Jupiter's atmosphere. It has become clear to planetary atmospheric scientists that the inner planets—the Earth, Venus, and Mars—formed atmospheres much different from those of the outer gas giants like Jupiter, Saturn, Uranus, and Neptune. Today, researchers have strong evidence to believe that the early atmospheres of the inner planets were comprised of carbon dioxide, carbon monoxide, nitrogen, and water molecules.

Atmospheric scientists now believe that any primitive atmosphere of reduced gases in the presence of water vapor would have rapidly oxidized through the process of photodissociation—the interaction of ultraviolet light rays from the Sun splitting apart water molecules to form hydroxyl radicals. In this scenario, ammonia and methane would be converted to nitrogen and carbon dioxide. Where did all the water vapor come from? Comets and other large bodies of ice and dust left over from the solar nebula rained-in on the inner planets in an epoch known as the bombardment period, which ended 3.8 billion years ago.

Venus, Earth, and Mars share similar amounts of carbon dioxide, indicating a similar origin for these worlds, even though much of Mars' carbon dioxide mysteriously emanates from the soil and some from the south polar ice cap. Much of our planet's carbon dioxide has since been absorbed in carbonate rocks through the dissolving of carbon dioxide in large bodies of water, such as the oceans. Of course, the primitive life forms on our world had an enormous impact on the production of carbonates and ultimately the carbon dioxide in Earth's atmosphere. We will explore the role life plays in the replenishment of Earth's atmosphere and how microorganisms have recently become recognized as a geological force.

Mars has a reservoir of carbon dioxide in the south polar ice cap

—but not nearly enough to account for its carbon dioxide atmosphere, which contains about seven millibars on average. Mars still has water in the form of permafrost, water vapor, ice-soil layer, and at times, liquid water. Evidence of carbonate rocks on Mars is demonstrated by the recent results obtained by NASA scientists studying the content of the SNC meteorites.

Venus, on the other hand, still has a massive carbon dioxide atmosphere (95-bar) that completely obscures the surface of the planet. Where does it come from? One theory is that carbon monoxide released through ancient volcanism became oxidized by water brought in from comets impacting Venus during the bombardment period. Since Venus has approximately the same mass of the Earth, its gravity would have prevented a primordial carbon dioxide atmosphere from escaping to outer space. However, early in its history Venus may have had lakes and oceans and even life. If so, carbonate rocks would have been produced by the same processes as on the Earth and Mars—life! Thus, the carbon dioxide content of Venus is thought to be ancient, whereas the carbon dioxide in the atmospheres of Earth and Mars is constantly recycled—but by what? Microorganisms.

The question that scientists concerned with the origin of life are now asking is: if life did not emerge chemically from a reducing atmosphere of hydrogen, ammonia, and methane, and if the neutral atmosphere of carbon dioxide and water vapor is the correct model, then how did organic material form and life emerge? It is interesting to note that there is not any "evidence" at all for the emergence of life, only the similarity of chemical composition.

Perhaps the most agonizing of all the questions centering on the Oparin-Haldane hypothesis of life on Earth is: why did not more than one biochemical form of life appear? Or did it, and all traces of its existence were obliterated by the "oxygen holocaust"? Perhaps, as Oparin once stated, any other biochemical form that may have gotten started was "eaten" by the dominant or already existing form. Since Mars apparently did not evolve to the point where photosynthetic life converted its atmosphere to oxygen, the life there could yet be of the ancient variety that was on the early Earth, before the oxygen holocaust. Studying any life that has evolved on Mars—and there is strong evidence it was indeed found by *Viking* in 1976—could lead exobiologists to answering the most-asked of all questions: What were the first living organisms like? How did life begin?

New lines of thinking promoted by the early experimental work of Harold C. Urey and Stanley Miller today include the continuing and revolutionary work being done by biologist Carl R. Woese from the University of Illinois. Woese, through his studies of polymerization involving dehydration linkages with monomers (a subunit of a polymer), has proposed that life on Earth began in the hot and dense atmosphere of the early Earth, as opposed to the Oparin-Haldane warm "biochemical soup" ancient Earth ocean scenario.

The most startling evidence of how life may have developed on ancient Earth has come from the extraordinary work of University of Miami biologist Sidney W. Fox. Fox discovered that the heating of dry amino acids can produce protein molecules, and that when these come into contact with water they form what he calls "proteinoids," which almost immediately upon getting wet take on the shape of microbial spheres that seem to "grow" by attaching themselves to one another and dividing! The spherical shape and membraneous appearance of the "proteinoids" are nearly identical to many forms of algae and bacteria, but nothing developed inside the tiny bubbles. The next challenge facing Fox and other biologists is to get from the proteinoids to something that could be called "alive." Getting a group of researchers to agree on "what life is" may be the most formidable challenge of all.

We have briefly reviewed here the classical chemical evolution theory of life on Earth, but what about Mars? The valley network channels on the cratered surface of Mars imply that Mars once had liquid water and a climate warm enough for water to flow. One of the interesting studies done by Dr. Daniel P. Whitmore of the University of Southwestern Louisiana's Department of Physics is detailed in a scientific paper for the American Geophysical Union's *Journal of Geophysical Research* (Vol. 100, pages 5457–64), entitled "A Slightly More Massive Sun As An Explanation For Warm Temperatures On Early Mars." Whitmore suggests that our main sequence star, the Sun, was "somewhat more massive" 3.8 billion years ago. If true, it would make sense that when the early solar system cloud of dust and gas condensed to form planets, asteroids, and comets that the energy from this slightly more massive Sun would be more intense in the outer regions around Mars and Jupiter than it is today. In this scenario it is entirely possible that conditions for life may have even arisen on Mars first, and then as the Sun began to lose its mass to solar wind expulsion and to condense 3.8 billion years ago, Mars started to become the colder world we know it as today. Taking this thought one step further, it is very possible that the seeds of life from an early Mars might have become implanted on Earth through the exchange of debris in the solar system 3.8 billion years ago, and in effect all of us could be Martians.

In a 1994 Symposium on Exobiology and Mars, Dr. James F. Kasting of Penn State University presented his abstract on "Habitable Zones Around Main Sequence Stars" and demonstrated that the zone of habitability, or HZ, of our Sun extends out to 1.37 Astronomical Units and perhaps "considerably further." Kasting defines the Habitable Zone by the point in orbit around a sun where planets have a "loss of water on the inner edge, and condensation of carbon dioxide clouds on the outer edge." This model puts the planet Mars just within the edge of the HZ today. The fact that Mars has water vapor clouds, condensate water ice or snow, and a dynamic atmosphere with seasonal changes similar to Earth seems to imply this is a world that according to a general theory of biology should have life.

Perhaps life on Mars could consist of the same type of early organisms that evolved on our planet billions of years ago, but because of the colder, more hostile climate of Mars, these organisms were prevented from evolving to higher forms of life. It would mean that life-bearing planets could be divided into two types, so far: one where life has taken significant control of the whole global chemical and physical environment, such as the Earth; and the second type, where life is at the mercy of the external environment—not capable of controlling it, just barely hanging on. This is why we must explore Mars while maintaining strict sterilization measures and quarantine control, for it may hold the ancient secret of how life began on Earth. Or perhaps, in the words of the great microbiologist Carl R. Woese, we explore Mars to find the so-called "Universal Ancestor." There is strong evidence today that life was detected on Mars by the Viking biology instruments in spite of the ongoing arguments against it, as we will explore in this volume.

Within the last ten years much has been learned about biology as new types of microorganisms are discovered here on Earth, while other fossilized evidence pushes the emergence of life back to the late bombardment period of our solar system, 3.8 billion years ago. How could life have begun so quickly in such an inhospitable world? The fact that fossil evidence for this ancient life seems to indicate that it was already in an advanced stage at this point brings into question the very idea that life even arose on Earth at all. Microbiologists have just begun to realize that as much as 99 percent of all microbial life on Earth has yet to be identified. These and other discoveries are literally causing a revolution in microbiology and related fields. Had scientists known of this at the time of Viking, they would not have been so hasty to preclude extant life on Mars. Now that organic material has been found and analyzed in interplanetary and interstellar dust particles, it raises even more fundamental questions regarding the origin of life. Does life originate in and around the gases surrounding newborn stars and then become implanted on all planets? Planets within the Habitable Zone of most solar systems might all, depending on their atmospheric components, be capable of sustaining life for at least brief periods of geologic time.

The theory that life originates in interstellar space and is deposited on the planets (the Panspermia theory) has been scoffed at by many leading astronomers. Then again, many prominent scientists proclaiming evidence for scientific discovery have been ridiculed by their peers. This has been a fact of our world since Copernicus, Bruno, Galileo, and Leeuwenhoek. We may have to conclude that a major scientific discovery cannot be made with a peer review because whenever that discovery strays away from the center of knowledge currently supported by those peers, they may have to admit they were wrong or otherwise unobservant.

Panspermia today has a somewhat different meaning than it did in the late nineteenth century. Individuals who upheld the Panspermia hypothesis in the last century thought that life could only come from life, not from inanimate matter. Today, Panspermia is generally defined as the transportation (not creation) of life from body to body in space. It says nothing about the origin of life and implies that life can grow anywhere conditions will permit it, whether in the interstellar dust clouds of our galaxy, the interior of comets, or the nebulous material that surrounds a newborn star and exists in the lithospheres of volcanic planets.

In 1971, a scientific paper entitled "Photocatalytic Production of Organic Compounds from Carbon Monoxide and Water in a Simulated Martian Atmosphere" (published in the March 1971 issue of the *Proceedings of the National Academy of Sciences)* demonstrated how a neutral atmosphere like that on Mars could produce substantial organic compounds. The significance of this discovery was enormous, and like the experiments of Miller-Urey, it proved that organic material can be created a number of different ways. To give this experiment a twist of irony, it was in part conducted by Dr. Norman H. Horowitz, the inventor and experimenter of the Viking biology experiment known as the Pyrolytic Release. The irony comes from the fact that even though Horowitz at first thought his experiment on the surface of Mars indicated a positive reaction, it did not indicate a positive reaction for life, only for the formation of organics. He measured the statistical significance of the wrong thing. In the final analysis Horowitz would say that life on Mars is impossible because the planet lacks liquid water.

Horowitz's 1971 experiment on Earth showed that when a mixture of carbon monoxide, water vapor, and carbon dioxide (all present on Mars) is irradiated with ultraviolet light, organic compounds such as acetaldehyde, formaldehyde, and glycolic acid are formed. Horowitz and his colleagues concluded in their paper that ultraviolet light reaching the surface of Mars would produce large quantities of organic matter over geologic time. It is interesting to note that a Harvard Observatory astronomer, William M. Sinton, discovered evidence for the organic aldehydes in the dark areas of Mars in 1957. This subject is explored in detail in Chapter Two of this book, "Dawn of the Astrobiologists."

For those readers who are not familiar with the twin *Viking* space-craft that were flown to Mars in 1975, it is necessary for you to know why they were sent—to search for life. When NASA announced in 1969 that it was taking scientific proposals to be considered in an automated mission to soft-land on Mars, it received an overwhelming 164 proposals, seventeen of which focused on the search for life on Mars. Of the seventeen life-detection experiment proposals, only three would get to test the soil of Mars—the Gas Exchange experiment (GEx), with its Principal Investigator, Vance I. Oyama, representing the NASA Ames Research Center; the Pyrolytic Release experiment (PR), with its Principal Investigator, Dr. Norman Horowitz, from the California Institute of Technology's Division of Biology; and the Labeled Release experiment (LR), with its Principal Investigator, Dr. Gilbert V. Levin, President of Biospherics Incorporated of Beltsville, Maryland. It is interesting to note that a fourth Viking Lander experiment called the Gas Chromatograph-Mass Spectrometer (GCMS), with Principal Investigator Dr. Klaus Biemann of the Massachusetts Institute of Technology's Department of Chemistry, would "politically" render the verdict; that in spite of the positive results obtained by the biology experiments, "no evidence" for life was found by Viking! This latter experiment was designed to look for organic material only—not life. We will examine in detail why the GCMS was used to exclude the possibility of life on Mars in Chapter Four.

Of course, the Viking Mission had much to say about conditions on Mars, including the geology, meteorology, and chemistry that would contribute heavily to those areas of science and if looked at closely, describe a planet capable of sustaining life. *Viking Lander 1* set down on the surface of Mars at 4:53 A.M. Pacific Daylight Time (PDT) July 20, 1976, seven years from the first manned landing on the Moon July 20, 1969. It landed in the northern hemisphere at 22 degrees north latitude in a region thought to be located at the mouth of an enormous river area (extinct). On a map of Mars this location is known as Chryse Planitia (Latin for "a level plain"). *Viking Lander 2* set down on the surface of Mars on September 3, 1976, in a more northerly location at 48 degrees north latitude in a low-lying plain called Utopia

Planitia. Utopia Planitia, because of its closer proximity to the north polar ice cap, was thought possibly to have more water in the soil either as a liquid or frozen just beneath the surface (which it did). It would not be until a few years later that geologists looking at the Viking photographs would speculate that Utopia Planitia might be the bottom of an ancient ocean basin thought to have covered the region millions of years ago with a depth of perhaps 500 meters.

As mentioned above, the Viking biology package contained three detection instruments for possible life on Mars. The Gas Exchange experiment (GEx) was to look at gas changes inside a small chamber where a portion of Mars soil was first exposed to water vapor only. Next, this sample was to be exposed to a liquid nutrient containing many different sugars, nineteen amino acids, organic compounds, vitamins, and inorganic salts. It was so complex the Viking Biology Team called the strange mixture "chicken soup." The GEx tested the humid-mode concept of life on Mars three times by offering water vapor only, and then attempted to promote some activity on the part of the microorganisms that might be present. Then it tested a wet-mode concept of life by "dumping" the "chicken soup" on the sample in hopes that microorganisms would react with one or more of the nutrients. Since this experiment was conducted in complete darkness, it was not looking for photosynthetic organisms.

When the GEx humidified the Martian soil (even before the "chicken soup" was applied), a large pulse of oxygen was released over a three-sol period (a sol is one Martian day). There was also a slow, steady release of carbon dioxide, peaking after two sols. However, when samples that had been heat-sterilized at 145 degrees Celsius for 3.5 hours exhibited the same reactivity, a biological explanation was rendered unlikely. Oyama and others speculated that peroxides, superoxides, or ozonides might have caused the evolution of oxygen in this experiment. But even if that were true, would not organisms adapted to the Martian environment be protected and able to exploit the oxygen contained in the soil? The answer to this question would take years of research on superoxides and hydrogen peroxide with terrestrial organisms.

The Pyrolytic Release experiment (PR), performed nine times

on Mars, looked for photosynthetic organisms in a soil sample. It supplied radioactive carbon dioxide and radioactive carbon monoxide (both present in the Martian atmosphere) to a sample of Mars soil exposed to a six-watt high-pressure xenon lamp mounted above the soil chamber. The idea was that if there were any photosynthetic organisms present, they would remove some of the atmosphere in the test chamber and incorporate it into the organisms. The excess radioactive gas was then removed from the chamber. The soil was heated to 635 degrees Celsius to "release" any radioactive gas that had been "fixed" by these organisms. If radioactive gas came off the soil sample at this time it would indicate that it had been previously incorporated into living microorganisms. The results of this experiment, although at first thought to be marginally positive, were in the end thought statistically insignificant as an indication of life.

The Labeled Release experiment (LR), which was carried out a total of nine times, used radiorespirometry microbial metabolism technology developed by Dr. Gilbert V. Levin. It supplied a dilute solution of five nutrients tagged with radioactive carbon (substrates) to Martian soil samples. No extra vitamins or other additives were included because the LR team felt all of those would already be in the soil since microorganisms (if there) would be growing in it. This experiment was carried out in the dark and initiated by releasing one drop of the radioactive nutrient solution on the soil sample inside the LR chamber. Water vapor is immediately available to all parts of the soil, with the center of the soil sample getting "wet" first, and as the water-substrate solution moves out over the soil, more of it becomes moist and then wet so that microorganisms are gradually subject to different conditions of moisture ranging from humid to wet. The LR was not looking for photosynthetic organisms, as the test chamber was in the dark. The LR results from the *Viking 1* and *2* landing sites, some 4,000 miles apart, were strongly positive and strikingly similar to tests on Earth. Heat was then applied to the sample to see how it would react, and at 160 degrees Celsius, the reaction was eliminated. This pattern had been seen many times with the LR preflight tests on Earth organisms. Taken alone, the LR results were a strong positive indication that life was discovered on Mars.

13

Because it took nearly twenty minutes at the speed of light for the first pictures of the Martian surface to reach Earth, none of the scientists could know if the landing was successful. The suspense was enormous, the stakes were high—after all, about *972.5 million dollars* (for two *Viking* spacecraft) was spent, with considerable cost attributed to the Lander biology instruments and Gas Chromatograph-Mass Spectrometer package, along with guidance, control, and sequencing computers capable of carrying out commands from up to 250 million miles away. It was the most ambitious scientific effort in the history of science—more complex than even the Apollo manned Moon landings.

Of the two *Viking* spacecraft sent to Mars, when the *Viking 1* and 2 landers separated from their orbiters, they became four spacecraft, two in orbit around the planet, and two on the surface of Mars. The orbiter missions were terminated when they ran out of fuel, with *Orbiter 2* lasting until July 25, 1978, and *Orbiter 1* until August 7, 1980. *Viking Lander 2* was shut down on April 11, 1980, because its batteries were depleted. *Lander 1* continued to relay information until November 13, 1982.

There is an old saying that makes the point "be careful what you wish for—because you just might get that wish." In the case of the Viking search for life on Mars, scientists got much more than they wished for, challenging the very foundation of their beloved Oparin-Haldane scenario of carbon-based life. For some, this may have been too much to comprehend at the time. The significance of the "strange" reactions in the soil of Mars would be re-analyzed by the same NASA scientists over and over, only to keep coming up with the same conclusion—"no life"—by most of them. Why?

In this book I will demonstrate that new knowledge about microorganisms and the enzymes they produce is the key to understanding Mars as a living planet. These enzymes can drastically alter rocks and soil, causing biomineralization—a process involving both biological and geological mechanisms in terrestrial desert environments. Biomineralization is a process most geologists considered to be theory just ten years ago. Since then, scientists working in the new field of geomicrobiology have been finding microbes everywhere on our planet—in the deepest oceans, on the highest mountains, in the driest deserts, on glaciers, and even living inside rock deep in the Earth. It is rapidly changing our perceptions of the origin of life. Of particular interest is how that information fits in with what we now know about Mars. Also, new observations being made by astronomers of organic molecules in the most remote regions of space may very soon paint a completely new picture of life in the universe, with the whole cosmos buzzing with life in every direction.

Mars: The Living Planet will demonstrate that Mars has all the right conditions necessary for life: water; water-ice; water vapor; ground fog; a soil with a heavy sulfur-iron content in which some terrestrial microorganisms could easily thrive; temperatures within the zone of habitability. There is evidence that Martian rocks and soil undergo interesting color changes that vary with the seasons, with many of the rocks having colored spots ranging from ocher to yellow, and even green. Why should color be an issue? NASA has maintained for over twenty years that Mars is red. Today, much of what is accepted about the surface properties of Mars is based on this simple observation.

Mars: The Living Planet is not intended as a review of previously published articles contained in books on the Mars Viking project or a recycled version of such, as most of those publications state that "no evidence" was found for life on Mars. Indeed, many of the scientists not willing to concede that the LR found evidence for life on Mars obviously do not know the definition of the word "evidence." Evidence is defined as being distinct from absolute proof, in that evidence is something to be considered when trying to determine whether proof exists. This book maintains the position that life on Mars was more likely discovered by Viking than not, and explains why—based on evidence.

Chapters One and Two of this book provide interesting historical background on how Mars became intuitively and later scientifically recognized as a planet capable of sustaining life by the early telescopic observers, and how the idea evolved into the twentieth century. As with any novel scientific departure, it is important to understand the historical context in which it arises because historical analysis always helps eliminate the negative impact of the status quo.

My decision to trace significant developments in the field of microbiology along with discoveries made by astronomers was based on the unique relationship the sciences share in their interpretation of the origin of life on Earth and in the universe. Astronomers of the late nineteenth and early twentieth centuries were primarily observers of phenomena explained in terms of color and shape. Remote sensing from Earth done with infrared spectroscopy equipment would not be useful until after World War II, and even at this point astronomers rarely had a substantial background in microbiology. But these two sciences would eventually merge for the greatest undertaking in the history of science—the Viking missions to Mars in search of evidence for life there.

The reader will find that I have interwoven the histories of astronomy and microbiology with important discoveries made by the *Viking Lander.* I did this to make a connection between important early milestones of microbiology and how they apply to the issue of life on Mars today. Also, I thought that by the time I introduced Dr. Gilbert V. Levin and the story of the Viking biology experiments in Chapters Three and Four, readers would have a working knowledge of what the Viking Mission to Mars was all about—the search for life. All the elements of each chapter pertaining to exobiology will be summed up in Chapter Eight, "ALH 84001 and a Martian Microbial Ecosystem." I hope this new approach in writing this book is both interesting and informative.

Subsequent chapters cover and explain scientific research in Antarctica and the role that harsh land has played with regard to life on Mars; examine the sensitivity of the Viking GCMS experiment that in the end rendered the "politically" decided negative verdict of "no life" on Mars for all three of the Viking biology experiments; and speculate that it may have been the GCMS that was in error.

Mars: The Living Planet will also look at the important issue of planetary quarantine procedures and the problem of forward and backward contamination of the Earth and Mars by organisms from each, and what the consequences could be from a Mars Sample Return Mission.

Some of the missions that are scheduled to fly to Mars between 1996 and 2005 are discussed with a perspective that could only be gained by someone who worked with NASA on its first orbiting spacecraft to Mars—*Mariner 9* in 1971, then *Viking* in 1976, and the Russian *MARS-96*—Dr. Gilbert V. Levin. Special emphasis is placed on the ill-fated Russian MARS-96 Mission, as this spacecraft would hold the only opportunity in this century to resolve the Mars life issue. But as you will read in this book, how a life-detection experiment found its way aboard the MARS-96 Mission in the first place is a very strange story that paints an interesting but sad picture of today's NASA.

Note also that none of the NASA spacecraft scheduled to visit Mars from 1997 to 2005 are planned to be equipped with a life-detection experiment, in spite of numerous efforts by some scientists to do so. This is indeed very unusual, considering that one of NASA's top three stated goals is to resolve the issue of life on Mars.

Just how difficult would it be today for a spacefaring nation to secretly return a sample from Mars by robot spacecraft if desired? A 1996 study by the Space Studies Board of the National Research Council placed an estimate of $300 to $400 million for a complete Mars Sample Return Mission. This includes a Lander with a sampler arm and Sample Return rocket that could return approximately four pounds of rock and soil directly to Earth. When compared to the $850 million price tag of NASA's Mars Observer Mission that was lost in space in 1993, this is a bargain. After all, *Mars Observer* was only a single orbiter. Also, it is difficult to understand how a massive project such as Viking, with two orbiters and two landers, could cost about the same as the relatively simple *Mars Observer*, even considering inflation as a factor.

In 1970, 1972, and 1976 the former Soviet Union demonstrated the ease with which a soil sample return mission could be accomplished with their unmanned *Luna 16, 20* and *24* spacecraft. Each returned a sample of lunar soil directly to the Earth. Could the same scenario be possible for a return sample from Mars? For all those individuals who believe that NASA would never do anything irresponsible regarding possible biological contamination from another world, they should beware the lessons of Apollo. During the manned lunar missions, no one knew for sure if the Moon did not harbor some form of lyophilized organisms protected under the lunar soil. The Apollo spacecraft splash-downs in the Pacific Ocean, even in spite of all best efforts by the crew to de-contaminate the cabin of their Command Module, were a terrible risk to Earth's biosphere. If Navy divers upon opening the Apollo crew hatch had released just one viable alien organism on a lunar dust particle that fell into the sea, then replicated and grew, it could have completely re-written the history of life on Earth! In other words, any alien organisms finding this new world an easy environment could, based on scientific studies (covered in Chapter Three), completely overrun our Earth in as little as six months or less. Apollo was a calculated risk on the part of the NASA science community which did not include informing ordinary American people of the facts.

Mars is not the Moon! It has an atmosphere of carbon dioxide, water, ice, clouds, and evidence for geothermal activity. These elements alone suggest that terrestrial microbial life could flourish there.

Finally, I have written this book for people who have pondered the question of extraterrestrial life in the universe—whether microbial or otherwise —and for those who have always questioned how the Viking biology experiment results yielded positive indications for life and then were later interpreted as a case of exotic soil chemistry. This book is by no means a definitive history on the subject of exobiology, and I encourage readers to go out and read whatever they can on this amazing new field of science in scientific journals located at their local college and university libraries. Many college libraries have computer access to the Internet as well, and there is an enormous amount of information available from all the NASA research centers, university research departments, museums, scientists, and more. This information field is growing at an incredible rate, and I suggest readers get on-line and "surf the web" for the latest scientific discoveries.

My goals for this book? I simply want as many people as I can reach to know that there is another side to the Mars life issue, and that one of the three Principal Investigators of the Viking Biology experiment package sent to Mars in 1975 says his experiment detected something in the soil of Mars that reacts just like microorganisms in soils on Earth. Furthermore, this scientist along with others has proposed other life-detection experiments that would settle once and for all the Mars life issue, but up to this writing have been rejected by NASA. Dr. Gilbert V. Levin's latest proposal for NASA and the *Mars Surveyor '98* program (with a lander) would not have cost the space agency much at all, but rather relied on technology that was already selected to fly on board. NASA's answer up to this point was still no. All thinking people should ask, why not?

Also, there seems to be a movement within NASA and certain factions of the scientific community to prevent Levin's and others' current scientific writings on the Mars life issue from getting into mainstream science journals and publications. NASA meanwhile expresses the view that nothing more than oxidants or peroxides was discovered on Mars, and anyone who opposes this view is looked upon as eccentric or irrational. Levin's list of scientific achievements, along with his patents and awards, paints a picture of innovation and success, not failure. Perhaps this is why some of his peers have even "borrowed" his ideas, then called them exclusively their own. To this end I have asked Dr. Levin to write the final chapter of this book, Chapter Nine, "Life After Viking: The Evidence Mounts." Dr. Levin's comments and observations are sure to get some attention!

Another special guest writer who is providing the final closing comments for *Mars, The Living Planet* is Dr. Patricia Ann Straat, who was Co-experimenter on the Viking Labeled Release biology experiment. Dr. Straat shared the tremendous moment of excitement at the Jet Propulsion Laboratory with Dr. Levin as the data indicating a positive reaction from the surface of Mars came in on the teleprinter. She also shared with Levin the negative outpouring from the scientific community.

The question today is, if life was discovered on Mars in 1976, then why would NASA keep this information to themselves? Why is NASA so concerned with bringing back a robotic sample of Martian soil to the Earth by the year 2003, before they make certain that the soil there has nothing harmful in it? After all, they have samples of

Mars rocks right here on Earth—in the form of the SNC meteorites that researchers claim are pieces of Mars blasted off its surface by an asteroid or large meteorite. Or is it more likely that NASA now knows that primitive life forms were discovered in 1976 by the Viking Labeled Release experiment, and they want to study these organisms secretly by bringing them to Earth as "sterile rocks"?

With more of the universe's secrets being revealed on almost a daily basis, we are on the threshold of new belief-shattering discoveries that will surely shake up society. A new era of truth and growth, perhaps?

CHAPTER ONE

Long Road to a Living Planet

The center of the Earth is not the center of the universe....
—NICOLAUS COPERNICUS, Commentariolus (1533)

M ars, the mysterious red planet with a timeless influence on the subconscious mind of humanity, has exerted its presence in many forms over the ages and even today calls to us from millions of miles out in space. Though at least twelve spacecraft have visited the planet in our century, we are still uncertain of this world that in many ways resembles the Earth. This uncertainty centers on three hundred years of telescopic and spacecraft observations of atmospheric, geologic, and soil surface samples that have yielded evidence for a planet that now appears to be a dynamic living world. Yet our complicated modern science has for the most part rendered logic impotent. Here is a world with water vapor, liquid water (at times), volcanically active systems, and a life-producing carbon dioxide atmosphere capable of sustaining (anaerobic) terrestrial microorganisms, let alone any that might have evolved on Mars. That life was most likely detected by the *Viking Lander* biology experiments in 1976 and then rationalized away by scientists creating theories about superoxides, hydrogen peroxide, and smectite clays —for all of which not a trace has been detected.

Such dissembling on the part of scientists and government puts Mars in a situation similar to the time when the prevailing center of knowledge said that Earth was the geocentric center of the universe. It makes you wonder about a program like SETI, the Search for Extraterrestrial Intelligence, where we spend millions of dollars looking for radio signals from another part of our galaxy: if we do finally detect a signal from another civilization, will we then rationalize that it could not be so? Earth has been at the biological center of our universe ever since recorded history. Can it be that we are not so willing to give up our reign? Clearly Mars is a threat in terms of altering our perceptions about life in the universe, and to those who fear it may change their fundamental religious cosmology. There is nothing more rewarding or frightening than exploring a new world. It usually has dramatic consequences.

In order to tell the time on the face of a clock, our noses must not be pressed against the hour hand, but rather we should stand at a distance. To fully understand the nature of Mars, it is necessary to look over the ages and glimpse the total panorama.

To the ancient Babylonians in 800 b.c., Mars was revered as a deity named Nergal, the most feared among all their sky gods. Also known as the "fire-star" to that civilization, Nergal-Mars gained a terrible reputation for violence because of its mysterious motions in the heavens and for being the color they interpreted as "blood red." These qualities led the superstitious astrologers of the time to believe that when Mars grew brighter it became "angry," and this was a time for war or sacrifice; when it dimmed, the astrologers took this to mean Nergal was no longer angered and good fortune would prevail.

What these ancient astrologers did not know was that the planets were in orbit around the Sun. To them, the planets were simply other stars in the sky whose rapid motions among the fixed stars in the constellations gave them authority over the heavens, and for this reason, they were deemed Gods. Mercury, Venus, Mars, Jupiter, and Saturn —the planets visible to the naked eye—all were worshipped as gods and messengers among the early civilizations. Mars, however, was especially revered because of the association of its color with blood, even though other red stars in the sky such as Betelgeuse and Arcturus were known. But it was Mars' rapid motion among the stars and its variable brightness that also caused alarm. Of course, Mars would become invisible as it disappeared behind the Sun in its orbit, so to the ancients, this only made Mars more mysterious.

It is interesting to note that Mars was known to other cultures a world away by similar concepts. For the ancient Chinese, Mars was Ying-Huo, the "fire planet"; to Aztec and American Indian tribes, Mars was Huitzilopochtli, "destroyer of towns and killer of people." Of course, Ares was the "planet of vengeance" to the Greek civilization, from which the Romans would later take as their deity Mars, God of War, and build an entire empire around the Red Planet. Mars was worshipped second only to Jupiter, King of all the Roman Gods. Because of the sacred association of blood and life in all these cultures, Mars symbolized both life and death along with sacrificial ceremony. Mars was looked upon as very much alive to these people of antiquity, but for very different reasons than we do today.

Along with their thoughts of the planets being sky-gods, many ancient cultures shared similar concepts regarding the origin of life on Earth. In China it was believed that aphids and other insects are self-generated from the effects of heat and moisture. The sacred books of India tell tales of various beetles, flies, and parasites coming from dung and sweat. To the Babylonians, worms and other animals came from the mud of canals. The ancient Egyptians held a similar belief that frogs, toads, snakes, and mice were formed from the layer of silt left behind by the flooding of the Nile. With few exceptions, all the early ancient cultures believed in some form of autogenesis. For all of them, the spontaneous formation of living creatures from non-living matter was an established fact based on the mystical interpretation of nature. This widely held view of the origin of life followed cultures from the East to the West and continued from antiquity to the Middle Ages. In Shakespeare's classic tragedy *Antony and Cleopatra*, Lepidus says that "crocodiles in Egypt are bred from the ooze of the Nile in the warm southern sun." In fact, autogenesis would persist until 1862, when Louis Pasteur overthrew the concept.

A.D. 1607 to 1830

The first clues to the true nature of the planet Mars would have to wait until the invention of the telescope, often credited to Hans Lippershey, a Middleburg eye-glass maker, in the year 1607. He noted that holding two lenses a distance apart could make far-away objects seem closer. Lippershey's invention was used only as a novelty at first until an Italian mathematician by the name of Galileo Galilei first heard about it in May of 1609.

Galileo took Lippershey's idea a step further by making his own more powerful telescopes capable of magnifying twenty times. He turned them towards the Moon and Jupiter, thus becoming the first telescopic astronomer in history. Galileo saw that the planets were worlds unto themselves and was astounded to see in 1610 that Jupiter had its own system of three satellites.

In that same year Galileo was approached by a monk from the Roman clergy about his discovery of the satellites of Jupiter and told that it did not fit in with the doctrines of the Holy Scripture. Galileo then was advised to abandon his opinions. He was reminded by the Roman Catholic clergy that the Italian philosopher Giordano Bruno also presented "unholy ideas" opposed by the Church and was subsequently burned at the stake just eleven years prior to Galileo's discovery of three moons of Jupiter. Bruno had asserted before his death, "Innumerable suns exist; innumerable earths revolve about these suns in a manner similar to the way the five planets revolve around our sun, with living beings inhabiting these worlds." Eventually Galileo himself would be considered a danger to the Church and was sentenced to house arrest for the remainder of his life because of his support for the heliocentric nature of the Copernican theory.

An unheralded quantum leap had occurred, and despite the efforts by the Church to silence new scientific discoveries, the universe would never again be looked at the same way by thinking people.

MEANWHILE, AT ABOUT THE SAME time (1609), another Middleburg lens crafter named Zacharias Jansen designed and built the world's first compound microscope, an invention that would play a crucial role in establishing the science of microbiology. Jansen's microscope opened the door to the later discovery of what types of microscopic organisms inherited the primitive Earth. It was never even considered in Jansen's day that the microscope would indirectly lead to the discovery of life on another planet—after all, the microscope was only intended to look at the smaller objects of our world, not something as enormous as the solar system or universe at large.

WHEN GALILEO GALILEI TURNED HIS telescope and attention towards Mars, he noted that the red world showed a phase similar to a gibbous

moon. However, because of the relatively small size of Galileo's telescope, he could not detect any subtle details on Mars. The largest
telescopes at the time had a focal length of only three feet. This was
because the craft of lens-making was still in its infancy. Word of
Galileo's observations of the heavens spread to other parts of Europe
and many became attracted to this new science of observing the sky
with a viewing tube.

As the years wore on, opticians discovered that if they increased
the focal length of their lenses hundreds of times, it provided much
higher magnifications. This led to some telescope tubes becoming
up to two hundred feet long! These were the great "aerial" telescopes
that were operated by the first wave of telescopic explorers. One can
imagine the hardships encountered while viewing with these mon-
strous telescopes, which often had to be mounted on masts or poles,
sometimes ninety feet high. These early astronomers had to wait until
the wind was perfectly still to observe on a night of clear skies, and
it would take the work of many other assistants to control the giant
wooden tubes from wandering away from their viewing subjects. This
was done by using a complex system of ropes and pulleys. Yet despite
the awkwardness of having to work this way, the first useful observations of Mars were made in 1636 by an Italian lawyer-turned-
astronomer named Francesco Fontana. Even though Fontana's optical
system had serious flaws, it enabled him to see a spot on Mars, and
following that spot as it crossed the planet, he deduced that Mars
rotates on its axis similar to the Earth.

In 1659, the Dutch physicist Christiaan Huygens drew the first
recognizable surface feature of Mars that we know today as Syrtis
Major. He also penciled in one of the polar ice caps, although Huygens did not comment on the nature of the tiny white spot. As Huygens continued to make numerous drawings of Mars, he was able to
make a determination of the Red Planet's period of rotation, which
he noted was almost exactly the same as the Earth, twenty-four hours.
This diligent explorer was also the first to comment on the color of
Mars, as he reports in his manuscript "Kosmotheoros" (1798), which
translates as "Cosmic Theories": "I am apt to believe that the land
in Mars is of a blacker color than that of Jupiter or the moon, which

is the reason of his appearing of a copper color."

Huygens worked with an "aerial" telescope 123 feet in length that had a lens 7½ inches in diameter. He was one of the first astronomers to discard the heavy wooden tubes that enclosed the objective lens. Instead, Huygens mounted his lens high up on a 90-foot pole and held the eyepiece 123 feet away, resting his elbows on a wooden brace. He achieved focus by having an assistant located at the top of the pole with a lantern, shining it in the lens and down to his eyepiece and him 123 feet away! Huygens, incredibly enough, using this same tedious method was able to discover the rings of Saturn and its largest moon Titan. The age of serious telescopic exploration had begun.

Christiaan Huygens might be considered one of the first telescopic explorers to comment on exobiology. He logically assumed that the Moon could not have advanced life because he could not see any clouds or evidence of water on its surface—an advanced notion for his time, as many other observers thought the Moon to be a perfectly suitable habitat for life. But he did feel Mars and Venus might be able to sustain life if they had clouds, water, and air. The telescopic revolution of Huygens' day demonstrated that the planets were other worlds, and it was quite natural for the astronomers of the era to believe that all the planets were inhabited, though they did not yet possess the scientific means to discuss it. All their theories were based on conjecture.

WHILE THE NEXT GREAT STRIDES in telescopic observations would have to wait until the mid-eighteenth century with the refinement of the Newtonian reflecting telescope, another giant leap was taking place in the realm of microbiology. Anton van Leeuwenhoek, a linen draper from Holland and amateur lens grinder, began making some of the finest microscopes of his time around 1670. In his age, "autogenesis" or "spontaneous generation"—the theory that living animals and other things could suddenly develop from non-living matter—still prevailed from ancient times. An example of this theory in the mid-seventeenth century was the putrefaction of meat. As the meat decomposed, it was observed that maggots formed seemingly out of nowhere. It was even thought that by storing corn and a linen cloth in a jar one could produce mice!

As Leeuwenhoek developed his skill as a microscopist, he began to observe for the first time in history microorganisms. To Leeuwenhoek, who called them "little beasties" and "animicules," this microscopic world of living things could be the answer to the diseases that had plagued humans for centuries and the true nature of spontaneous generation. He tried desperately to get the Royal Society of London to review his work until his death in 1723. However, Leeuwenhoek's peers scoffed at his discoveries and merely said it was "exotic soil chemistry and not life" which he was observing through his microscope.

Had Leeuwenhoek encountered the same phenomenon that Ferdinand Magellan encountered at Tierra del Fuego? According to Magellan's journal, when his ships arrived in the bay of Tierra del Fuego, the natives who lived in the nearby villages were unable to discern the vessels. It was only later, when the native tribe shaman pointed them out, that the ships began to materialize in their minds as being real!

Does the human mind have a mechanism that blocks out information if that information does not fit our belief system? According to the journal of Magellan, it did. How could Leeuwenhoek's peers dismiss the moving microscopic matter under his microscope as exotic soil chemistry without exploring the discovery more closely? Could it be possible that scientists today, holding on to current concepts of a dead, sterile Mars, cannot "see" that other evidence exists for life, but like the ships in the harbor at Tierra del Fuego, it is beyond their belief systems? Already many scientists have abandoned the idea of life on Mars, even though the evidence for its existence is in plain view in the form of Viking data and new information regarding living organisms on Earth.

GERMAN-BORN SIR WILLIAM HERSCHEL left for England in 1757. Herschel, who had worked for twenty-five years as a professional musician and was accomplished in twelve different instruments, began to immerse himself in the study of optical design and construction. He did so with an appetite for astronomy and hoped to improve the

Newtonian reflecting telescope design originally built
by Sir Isaac Newton in 1671.
Newton had set out to
design a telescope that would
replace the giant aerial
refracting telescopes by fold-
ing the optical light path in
half using a mirror instead
of a glass lens.

As only inferior grades
of optical glass were available in Herschel's time, he
decided to abandon the
refracting telescope and pursued the shaping of optical
mirrors, which he found
were much simpler to make,
as they were fashioned from
a composite of copper and
tin and not brittle glass.
These mirrors or "specu-

Sir William Herschel discovered that Mars had seasons like the Earth, only twice as long.

lums" could be made very large because of their ease to work with,
and Herschel proceeded to make the largest telescopes of their kind
in the world. His biggest had a diameter of five feet for the mirror
and a length of forty feet for the tube. This giant telescope was
mounted on a giant rotating platform that could be positioned to
look anywhere along a meridian passing through the zenith.

Although Herschel is remembered most for his incredible undertaking of cataloging every star and nebula he could find, he also was
an avid Mars observer and the first to comment that the polar ice
caps of Mars might be snow and ice. He determined the Martian seasons, which to his delight were twice as long as Earth seasons and
therefore led him to conclude incorrectly that Mars must have a
milder climate than the Earth. Among his other Martian discoveries
was the fact that Mars had clouds and weather.

Herschel, like many of his contemporaries, also thought that all the planets contained life, even intelligent life. But when he announced his unusual theory of an inhabited Sun, he met resistance from his peers. Herschel's theory was that the Sun merely had a luminescent atmosphere hiding its surface, which he believed would be similar to the Earth and inhabited. In spite of this naive line of thinking, it was Herschel who by the time of his death in 1822 ushered in a new age of astronomical discovery that would accelerate feverishly to the present day. More importantly, he prompted other astronomers by means of his careful observations of Mars to examine this world of seasonal changes, melting polar ice, and wind-blown clouds.

1830 to 1916

While the giant reflecting telescopes of Herschel's design were still enjoying extensive use throughout the first half of the nineteenth century, the science of grinding precision lenses was coming of age and finally surpassed the optical images that could be obtained by large reflecting telescopes. While the large reflectors were fine for gathering light from distant star systems, it would be the refractor, with its finely ground and polished lenses, that would again revolutionize the astronomical kingdom.

One of the finest lens craftsmen of the eighteenth century was a German optician by the name of Joseph Fraunhofer, who became a legend in his own time for his outstanding achromatic lenses. Although Fraunhofer died in 1826, his pioneering work in the development of the refracting telescope would advance planetary observations and cartography through the rest of the century. Fraunhofer was also the first to use a prism system combined with a telescope to analyze and map the absorption lines of the solar spectrum, and in doing so created the stellar spectroscope. He discovered that the chemical elements of the Sun and other stars could be examined through the new instrument. The spectroscope would become the most powerful astronomical tool for analyzing the atmospheres of the planets, although it took until 1947 to perfect its use and obtain definitive results. The spectroscope would later be incorporated into every spaceprobe sent into space because of its versatile ability to analyze the chemical elements in the atmospheres of planets and stars.

German astronomers Wilhelm Beer and Johann H. Madler would use a small 3¼-inch diameter lens made by Fraunhofer to make the first crude map of Mars during the years 1830 and 1832. At their Berlin observatory in 1837, Beer and Madler used this telescope to make their "great map" of the Moon, which had such stunning detail that a new wave of astronomers wanting only refracting telescopes began to emerge. It was soon realized that a small refractor was able to render the better contrast and details needed for planetary studies than a reflector many times its size.

ANOTHER SIGNIFICANT EVENT UNFOLDING IN 1856 centered on the writings of a young seafaring naturalist who had previously spent several years at sea on a British ship, the H.M.S. *Beagle*, touring the world's most exotic locales, including New Zealand, Australia, the Canary Islands, and South America. A thirty-seven-year-old Charles Darwin developed his observations and theories based on his investigations and travels. He originally planned to write a large six-volume monograph entitled *Natural Selection* which would detail his ground-breaking theories of evolution.

One day in 1858, Darwin stopped work on this massive project because of a letter he received from another young naturalist working in Malaya. Apparently the writer from Malaya had formulated his own theory of evolution and threatened to scoop Darwin on his life's work. Darwin immediately abandoned the larger work and broke down the information it contained thus far to present a finished abstract entitled "Origin of the Species" in 1858. The book completely sold out on its first day of release. Along with his famous theory of evolution which predicted all living things could be traced through "reductionism" and "synthesism" to a common ancient universal ancestor, Darwin also defined the characteristics of living organisms: a) all living things maintain a complex organization; b) all living things use matter and energy for growth; c) all respond to stimulation from their environment; d) all living things adapt to changing environments; and e) all living things reproduce.

The impact that Darwin's work had on the fields of microbiology and paleontology would eventually lead directly to the first logical considerations for the evolution of life on other worlds, or the science of exobiology.

IN 1858, ANGELO SECCHI, AN Italian priest and astronomer and Director of the Gregorian University Observatory in Rome, gained notice for his work in the new science of spectroscopy. Secchi carefully cataloged the colors (spectra) of thousands of stars, which he did with only a 9½-inch refracting telescope. Secchi was also one of the first to apply the principles of spectroscopy to planetary observations, although his results were disappointing and refinement of the spectroscope for planetary observation would have to wait.

One of the other areas that most intrigued Secchi was the color of the planets, and in 1858 he rendered the first color drawings and charts of Mars. The dark areas of Mars were represented in shades of green, with the lighter zones shown in yellow. The green areas he began to suspect might be vegetation. Where was the ruddy red color of Mars in Secchi's drawings? Since he had trained his eye to the sensitive task of observing color in the spectrum of thousands of stars, Secchi only rendered what he saw, and it was not the color red.

In 1865, a French astronomer named Emmanuel Liais who worked out of the Paris Observatory published a book, *L'Espace céleste*, which contained detailed color drawings he had made of Mars in 1860 showing green in many areas. Liais concluded that the green areas must be vegetation. A virtual explosion of Mars fever was about to burst upon an unsuspecting world.

IT IS IMPORTANT TO POINT out at this time that the brilliant French chemist Louis Pasteur in 1864 finally managed to refute once and for all the concept of "spontaneous generation," as discussed earlier in my account of Leeuwenhoek. Pasteur achieved this by boiling infusions in his famous swan-neck flasks, which permitted the free access of air into the flask contents, but prevented microorganisms in the air from entering. The untreated air passed in and out of the flask but no microbial growth appeared in the liquid. This demonstration led to the practice of sterile technique and the prevention of many infections.

More importantly, and distantly relating to the *Viking* discovery of microbes on Mars, would be Pasteur's experiments with microbial decomposition. He found to his astonishment that there were microorganisms that could live without any oxygen. He called his newly discovered organisms "anaerobes." Recall from my discussion in the introduction of this book that the early Earth did not have an oxygen atmosphere and that the first organisms to evolve used the gases that were present from volcanic activity and other sources.

"Father" Angelo Secchi, an Italian Priest, was the first to make drawings of Mars in color.

DOROTHY SCHAUMBERG/MARY LEA SHANE ARCHIVES OF THE LICK OBSERVATORY

The explosion of scientific knowledge and theories in the disciplines of microbiology, biology, chemistry, and astronomy occurring during the mid-nineteenth century was nothing less than a complete revolution. So much was happening so fast that it was difficult to keep up with the latest discoveries made in these fields. As Pasteur was disproving the notion of spontaneous generation of life, a new concept was initiated by a German physicist, Hermann Richter, in 1865, who had an intense interest in meteorites. He hypothesized for the first time that "living germs" might be brought to the Earth by cosmic dust and meteorites.

Richter's idea was embraced by his contemporary Hermann von Helmholtz, who discovered the concept of the conservation of energy in 1854. Together with Sir William Thomson, who later became Lord Kelvin, Helmholtz added an idea to Richter's hypothesis: that germs together with space dust were transported through space by the action of light pressure from stars. Richter's theory was known as "litho-panspermy," a derivative of the Greek word "litho" meaning "rock," with "panspermy" translating as "everywhere" and "seed." Helmholtz and Thomson's version was called "radio-panspermy." In this new version of Richter's hypothesis Helmholtz wrote, "It seems to me that if all our attempts to create organisms from lifeless material are unsuccessful, it is scientifically reasonable to ask ourselves the question, Did life indeed ever originate, or is it as old as matter?"

The Panspermia hypothesis seemed to be a rational approach to the implantation of living organisms throughout the universe. For Richter, meteorites passing through the atmosphere were heated only on the outer surfaces (a fact true of small meteors today); thus the inside was presumably cooler and therefore any organisms in the meteorites were preserved. Upon hearing the theory of Panspermia, Louis Pasteur began to examine meteorite specimens for evidence of bacteria or fossils of extraterrestrial organisms. In 1880, Dr. Otto Hahn, a German physician and author, wrote a book entitled *The Chondrite Meteorites and Their Organisms*, in which he claimed to have discovered the fossil remains of tiny organisms. But he was later convinced by his peers that all he found were odd structures in the chondrites that resembled living mosses but were in fact just mineral deposits.

DURING THE MID-NINETEENTH CENTURY, Mars continued to be of great interest to astronomers because of its similarities to the Earth. Mars has a close approach to Earth in its orbit every two years called an "opposition." Each succeeding opposition yielded new information on the colors of Mars, its surface markings, and weather patterns. However, in 1877 Mars was brought to the forefront of the world's scientific theatre—quite innocently at first, and later to encompass a psychological phenomenon not repeated in scientific history:

the discovery of the Martian "canali" by Giovanni Virginio Schiaparelli, who ironically graduated with honors from the University of Turin in 1854 as a hydraulic engineer!

Schiaparelli became second astronomer at the Brera Observatory at Milan in 1860. He also was an accomplished Latin scholar and the first astronomer to discover the relationship between the orbits of comets and meteors in 1866, suggesting that meteors and meteorites are products of the "dissolution of comets." He noted that after the appearance of Comet III of 1862 a meteor shower ensued in August, and he repeated this unique observation for a comet and meteor shower in 1866.

Giovanni Schiaparelli was the first astronomer to conclude that the Sun repelled the particles in the tails of comets, although he did not fully comprehend the nature of the solar wind. This discovery, however, would lead other scientists including Svante Arrhenius to propose the theory that the rays of the Sun could move dust and small debris in interplanetary space, and in doing so, might deposit living "germs" attached to dust grains on the surface of planets. Hence the concept of Panspermia was given a plausible means.

Schiaparelli discovered the asteroid Hesperia in 1861 and devoted the later part of his life to studies of ancient Babylonian astronomy, with a particular emphasis on what that ancient civilization recorded in their observations of Mars and Venus. He was well acquainted with all the sacred books and principal religions and was deeply interested in the historical development of Christianity—a unique combination of intellectual talents for this master of life's curiosities.

Of course, Giovanni Schiaparelli's most famous discovery was the fine dark lines crisscrossing the disk of Mars. He termed these "canali," which translated from Italian merely meant "grooves." From the Royal Brera Observatory in Milan, Schiaparelli used only a modest $8\frac{3}{4}$-inch refracting telescope for his Mars studies. While conducting a trigonometrical survey (the first ever attempted) of the disc of Mars, which had 25 inches of arc size, he detected a new and curious feature. What had previously been thought to be Martian continents were found to be conglomerations of "islands" separated by a network of hair-thin lines with a gray-green hue and measured by

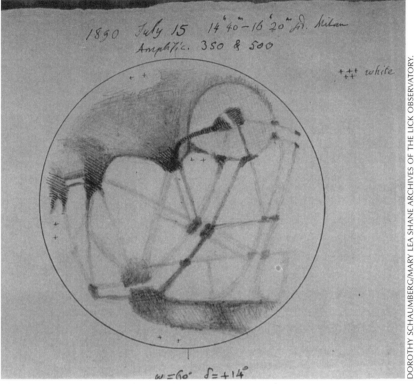

The "canals" of Mars as drawn by Giovanni Schiaparelli in 1890.

Schiaparelli to be three or four thousand miles long and sixty miles wide!

When word of his discovery hit the astronomical community, it spread like a brush fire. Could it be that other astronomers of the time were so unobservant that they completely missed this incredible phenomenon, or was it that the opposition of 1877 brought Mars closer to Earth than it had been in recent history? Whatever the reason, astronomers everywhere wondered what could cause such a peculiar phenomenon and they set out to investigate it.

While traveling the Far East as a successful businessman, a young Percival Lowell read of Schiaparelli's discovery of canals on Mars. Interested in astronomy since his youth, Lowell often took along his 6-inch reflecting telescope on business trips. As Lowell glimpsed what he thought were Schiaparelli's canals, an idea began to come to him that the canals might be of intelligent design. He felt so compelled and convinced by this concept that he set out to build his own observatory. Because Lowell and his family owned a textile empire, he could on a whim do almost whatever he wanted, but Lowell was also a determined eccentric who spent two years looking for the ideal spot to build his Mars observatory. After taking careful notes on several observing locations around the globe, Lowell settled for Flagstaff, Arizona, which in his opinion had the best "seeing" conditions. This is usually referred to by astronomers as a combination of clear and dry sky conditions.

A young Percival Lowell before his proclamation of intelligent life on Mars in 1875.

Lowell wasted no time in the construction of his observatory, and in 1894 made the request to "borrow" a large 18-inch refracting telescope from Harvard University, of which he was an influential graduate. Lowell along with fellow Harvard contemporaries A. E. Douglass and W. H. Pickering set up his famous observatory on "Mars Hill." In 1894 he compiled a series of 917 drawings of the planet showing in exquisite detail Schiaparelli's canals, and of course many newly discovered ones by Lowell.

When Percival Lowell began to voice his theory on the canals having an intelligent extraterrestrial origin, newspapers around the globe began to print it. Lowell's contention that a dying race of intelligent creatures were diverting water from the polar ice regions of

Mars to the equator became the tabloid sensation of the world. Thousands of interested individuals purchased telescopes for the primary purpose of seeing the canals of Mars. It turned to hysteria when amateur astronomers began reporting to their various astronomy clubs and groups that signals were being flashed from Mars to Earth by the civilization that Lowell said created the canals. These signals, as it turned out, were observations of real phenomena, only they were in fact clouds and possible dust storm activity with an anthropomorphic interpretation.

Lowell certainly accomplished one thing with his canal theory: it got everyone interested in Mars. Many of the professional astronomers of Lowell's time said they could see nothing of the canals that Schiaparelli or Lowell reported. One of the prominent artist-astronomers of the time was Nathaniel Everett Green from England, and in 1877 he made his own Mars map based on drawings created on the Island of Madeira, off the coast of Morocco. Green's map depicted the surface features of Mars as it is generally observed by today's seasoned amateur astronomers. Green used a 13-inch refracting telescope and rendered subtle yellows and oranges for the desert regions of Mars and greens for the dark areas. He also penciled in the white polar ice caps, but saw nothing that resembled canals! Upon seeing Green's Map of Mars, Schiaparelli felt compelled to write to him, insisting that the canals of Mars "are as definite as the rivers on the Earth." Green remained unchanged in his opinion and simply replied that his observations and drawings of Mars were correct. He later accused Schiaparelli and other astronomers drawing canals of "not drawing what they see." Those astronomers who claimed the canals were real said that those who could not see them had poor technique or bad eyesight.

To fuel this controversy even further, the respected French astronomer Camille Flammarion—who carried out meteorology experiments and measurements in Earth's lower atmosphere by ascending in hot-air balloon flights —supported Lowell's conviction of a race of intelligent beings on Mars, adding, "Yet we may hope that, because the world of Mars is older than ours, mankind there will be more advanced and wiser." Flammarion was editor of the French

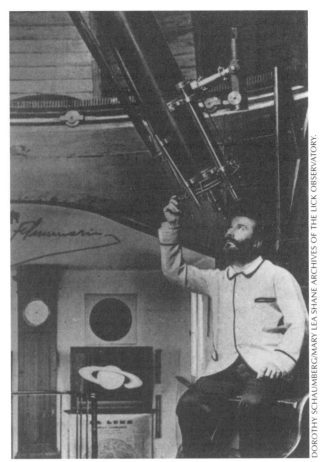

Camille Flammarion in his home observatory in France in the late 1800s. Flammarion fueled Lowell's concepts of intelligent life on Mars.

journal of astronomy, *L'Astronomié*, and went on to publish the most comprehensive books on Mars ever done: *La planete Mars*, Volumes 1 and 2 (1892 and 1909), each exceeding five hundred pages in length and detailing everything known about the planet at the time. Flammarion would go on to publish over twenty-five other volumes on astronomy and space. He remained loyal to Lowell's concept of an inhabited Mars throughout the remainder of his life.

In 1893 Giovanni Schiaparelli published a summary of observations he had made of Mars up to that time which described an even more peculiar aspect of the mysterious canals. At times they would appear to double right before his eyes! In the February 15, 1893, issue of *Natura ed Arte*, Schiaparelli says,

> Two lines follow very nearly the original canal, and end in the place where it ended. One of these is often superimposed as exactly upon the former line but it also happens that both the lines may occupy opposite sides of a former canal, and be located upon entirely new ground. The distance between the two lines differs in different "geminations," and varies from 600 kilometers and more, down to the smallest limit at which two lines may appear separated in large visual telescopes—an interval less than 50 kilometers.

Schiaparelli believed the lines and bands on Mars could be explained as the result of physical and chemical changes, but said that the seasonal color changes had to be related to organic events of enormous magnitude similar to the flowering of mountainsides on the Earth. He also felt that the canals or lines crossing Mars "don't teach us anything about the existence of intelligent beings on this planet."

In 1903, one of Lowell's most outstanding critics from England, E. Walter Maunder, published a book entitled *Are the Planets Inhabited?* In the book, Maunder gave an example of how the human eye and brain coordinate in conjunction with the observing of fine detail. Maunder gave his class of two hundred students at the Greenwich Hospital School in England instructions to copy a drawing of Mars he made and posted in the front of the classroom. The drawing represented Mars as he had seen it and drawn it, without any canals. To Maunder, the canals were merely minute details that were not resolved by small telescopes. As a demonstration, he had his students at the front and rear of the classroom render drawings of what they saw from where they sat. Maunder examined their finished drawings, noting that the students who were further away from his Mars drawing, in the rear of the classroom, drew lines resembling the canals, whereas those students in the front of the class rendered drawings with mottled details. When Lowell read Maunder's book he countered by stating that if one observes a telegraph wire at various distances, it will always appear as a straight line.

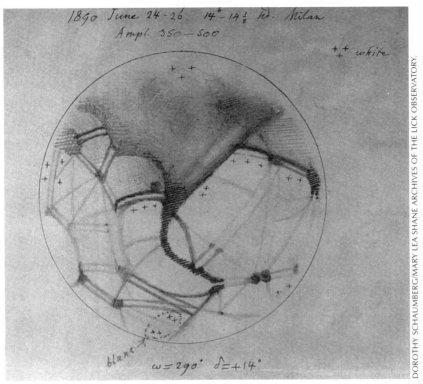

Another Schiaparelli drawing of Mars showing the doubling of canals, "geminations," another psychophysiological phenomenon associated with Mars in the late 1800s.

The canal phenomenon even captured the studious attention of the British Astronomical Association, where reports by members often exceeded the number of canals mapped by Schiaparelli or Lowell. Everyone, it seemed, wanted to discover a new canal. The debate over the canals continued until the arrival of the *Mariner 4* spacecraft in 1965. The twenty-two photographs returned by the spacecraft showed a planet devoid of any canals. However, because the canals are a result of minute details blending together at a distance to form straight lines, the phenomenon can still be seen in small telescopes today as a largely psychophysiological effect.

As important as the discovery of the canals seemed at the time, the significant contributions by nineteenth-century astronomers were

41

their observations of color changes, cloud formations, and limb hazes. Clouds observed on Mars were of three types: white clouds and morning fogs, blue limb hazes, and yellow clouds, interpreted as dust being swept up into the atmosphere. It was also observed that a haze enshrouded the north polar ice cap, which hence was named the north polar hood. As the phenomenon of the yellow dust clouds became better understood to represent dust storms on the planet, it seemed no one bothered to ask the obvious question: If the surface of Mars is pink, orange, or red as many astronomers even today have claimed, then why is dust from the surface of Mars yellow? The answer to this intriguing problem would have to wait until the *Viking* spacecraft touched down on the soil of Mars in 1976.

Whatever criticisms Percival Lowell had to suffer because of his belief that a race of intelligent creatures existed on Mars —a belief he would carry to his grave in 1916—he was a keen observer of subtle color changes on Mars. In fact Lowell's most lasting contribution, though it is still debated today, was his observations of the "wave of darkening," a surface, not atmospheric, phenomenon. It occurred when the polar ice caps on Mars began to melt, and as water vapor transported through the atmosphere interacted with the surface of the planet, it caused these areas to become progressively darker. The darkening continued from the polar region in a wave-like motion down to the equator and then into the opposite hemisphere. So precise was the appearance of this wave of darkening descending from the melting polar ice cap that its speed crossing the planet could be accurately measured by Lowell—about thirty kilometers per day.

Many of the astronomers during Lowell's day thought that the feature known as Syrtis Major (formally Hourglass Sea) was actually an ocean because of its deep blue-green hue. But Lowell had another theory, as quoted from an article written for *Popular Astronomy* (No. 36) in 1906:

> A sea [Syrtis Major] it continued to be by generations of astronomers. For over two hundred years its supposed aquatic character passed current practically unquestioned. Its color seemed in fact conclusive of its constitution... Unfortunately, for the ocean-loving character of Syrtis Major must pass with other

charming myths into the limbo of the past. For the great blue-green area is no ocean, no sea, no anything connected with water, but something very far removed from water: namely a vast track of vegetation ... the color of Syrtis Major in places fades out. Just such an effect would follow the change of vegetation from green to yellow and ocher, as spring passed toward autumn.

Lowell's theory of a wave of water vapor from the polar regions descending down past the equator while creating vast color changes

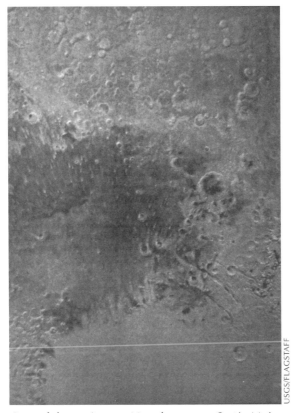

Part of the region on Mars known as Syrtis Major which, from Earth-based astronomers' observations, appeared to grow and change color (green and blue) in Martian spring-summer and gradually fade in the autumn-winter (to dull brown-gray).

was the first substantial evidence to indicate Mars had an ecosphere and could be a living planet. Like a ghost, Lowell's wave of darkening has followed astronomers of the early twentieth century right up to our present time.

ONLY THREE YEARS AFTER THE development of spectrum analysis, English astronomer William Huggins began his original work looking for traces of water vapor in the atmosphere of Mars using a two-prism spectroscope and a 5-inch refracting telescope from his home observatory in London. To Huggins, routine visual observing methods became boring, and the spectroscope offered a wealth of uncharted territory to cover: mapping the chemical elements of the Sun, Moon, and planets.

It was not until 1867 that Huggins got a result that required serious consideration. Because the Earth has an absorbing atmosphere, early spectroscopists like Huggins wanted to be sure that the spectrum of Mars was actually being viewed and not the spectrum of the Earth's atmosphere or the reflected solar spectrum. Since astronomers knew that the Moon did not possess an atmosphere, they compared all planetary spectra to the Moon's, and if there were any odd lines that did not match the solar spectrum or that of the Moon, it was then assumed that the spectra belonged to the planet being observed. Huggins found a line in the spectrum of Mars that did not belong to the Moon, Earth, or the Sun, although he did not know what the element in the spectrum was. The spectrum was still being explored and mapped, and the only real documented elements were those found in the Sun earlier by Fraunhofer and others. This definitely was uncharted territory.

With all the controversy surrounding Lowell and the canals with his discovery of the wave of darkening, it is not surprising that scientists at the time hoped to prove the existence of water vapor on Mars. Such a discovery would instantly make an astronomer famous and well respected among his peers. After all, here we have a planet going through seasonal changes with its landscape altering colors from greens in spring and summer to browns and grays in fall and winter. If there is life on Mars, then there had to be an abundance of water, it was reasoned. In 1867, Pierre Jules Cesar Janssen, director of the Astrophysical Observatory in Meudon, France, announced that he had discovered the presence of "aqueous vapor" in the atmospheres of Mars and Saturn. Even though Janssen had worked earlier with Huggins on his pioneering studies of the spectrum of Mars, Janssen's work was considered dubious, as he failed to officially publish his results with the required corresponding data.

In 1873, Professor Hermann K. Vogel claimed from spectroscopic studies he had made of Mars, "It is definitely settled that Mars has an atmosphere whose composition does not differ appreciably from ours; and, especially, that atmosphere must be rich in aqueous vapor." Vogel had the data in hand, both water vapor and oxygen were detected in his spectroscopic analysis, but was it of Mars? It was later discovered by W. W. Campbell of the Lick Observatory in California that Vogel had succeeded in finding water vapor in Earth's atmosphere! In 1894 Campbell made his own spectroscopic analysis of Mars and found no difference between the spectra of the Moon, a dead world, and Mars.

It was finally concluded that unless the atmosphere of Mars had a greater abundance of water vapor than that of Earth, it would be useless to look any further with the equipment being employed. It just was not sensitive enough to search for water molecules. However, Professor Lewis E. Jewell of Johns Hopkins University stated in Volume 2 of *The Astrophysical Journal* (1895):

> It seems to me, however, that there may be more hope of detecting the presence of chlorophyll than either oxygen or water vapor in the spectrum of Mars. Of course it will require careful and delicate observations, but the chlorophyll band in the red end of the spectrum of vegetation is quite strong; and if the green areas of Mars are due to vegetation, this band might be seen. In speaking about this matter with Professor William H. Pickering a few weeks ago, I asked him why no one had looked for this band, and he informed me that he had looked for it, but that the season on Mars was so far advanced at the time that the areas that had been of a strong green color had become gray at the time his observations were made, so they were not conclusive. [*See Color Plate 1.*]

This line of reasoning—that if the green areas of Mars were due to vegetation, they should reveal themselves in the chlorophyll band of the spectrum—led to the idea of comparing this spectrum of the green areas to the spectrum of the desert areas on Mars. It was a fantastic idea at the close of a century of remarkable discoveries. It would set the stage for the twentieth century.

CHAPTER TWO

Dawn of the Astrobiologists

Questions involving both astronomy and biology were posed for the first time in 1877 when Schiaparelli of Milan discovered the features on Mars named by him "canali."
—DR. HUBERTUS STRUHOLD, Chairman for the First American Symposium on Astrobiology, June 1957

B EFORE THE CLOSE OF THE nineteenth century the science of microbiology was invented by Pasteur's work refuting the prevalent theory of "spontaneous generation." He went on to prove that microorganisms are the source of biodegradation of organic substances. Several very significant discoveries were made by nineteenth-century microbiologists studying microbial ecology and the enzymes known as catalase and peroxidase, and this knowledge has important implications for the 1976 *Viking* spacecraft biology results if indeed Mars is eventually found to have a small amount of hydrogen peroxide in its soil (as predicted by some scientists). Proof of this theory would have to wait until the Russian Mars-96 Lander tested the soil of Mars in 1997 with a U.S. experiment called MOx (Mars Oxidant experiment) designed to look for chemical oxidants including peroxides. But launching spaceprobes is not without risk, and success can never be guaranteed, as discussed in Chapter Six.

In Earth's atmosphere, hydrogen peroxide is formed when the Sun's ultraviolet light breaks apart water molecules to form hydrogen (H) and hydroxide (OH) free radicals. Two OH radicals then combine to form hydrogen peroxide. Hydrogen peroxide has been found at various concentrations in cloudwater, rainwater, and soil. However, ultraviolet light also rapidly destroys hydrogen peroxide by turning it into water, oxygen, and ozone. Ozone, ironically, protects macroscopic living organisms on Earth by absorbing the harmful portions of the ultraviolet spectrum. If not for the ozone in our atmosphere, most forms of life would die from the intense ultraviolet radiation. A small amount of this radiation is what gives you a sunburn, so you can imagine what an exposure thousands of times greater might do to the human body. But microorganisms living on Earth before an ozone shield built up in the atmosphere developed many surprising ways to protect themselves, as we will discover.

It has been found that somewhat significantly higher hydrogen peroxide concentrations occur as a result of urban atmosphere pollution, and that hydrogen peroxide levels increase through oxidation of hydrocarbons (Zika and Saltzman, 1982). Hydrogen peroxide is found in the surface waters of the world's oceans, produced by photochemical reactions of sunlight and splitting of water molecules. An interesting but not well understood fact regarding measurements of hydrogen peroxide concentrations on the Earth comes from data gleaned from Greenland ice cores. These samples show that the global atmospheric level of hydrogen peroxide has increased by fifty percent during the past two hundred years (Sigg and Neftel). It is possible that hydrogen peroxide played a major role in the development of life on Earth and might still play a role on our planetary neighbor, Mars. To understand how this might be possible—along with other fascinating scenarios for early processes involving organisms and oxidation—we now take a brief tour of some discoveries in microbiology that were made in the nineteenth and early twentieth centuries. Information obtained by the Viking search for life on Mars might also shed light on an intriguing new concept about hydrogen peroxide and organisms that use it for oxygen and water.

In 1855, it was discovered that unique "substances," or enzymes, occurred in some plant and animal tissues that could decompose hydrogen peroxide into water and oxygen (Schonbein, C. F., 1855). C. Schonbein, who had discovered ozone in 1840 while conducting chemistry experiments, found that it would turn guaiacum blue. While guaiacum had been used since 1810 as a dentifrice and had been noted to turn blue in the mouth, it was Schonbein who demonstrated its usefulness as a test for oxidation ferments.

Hydrogen peroxide is a highly reactive and oxidizing chemical produced naturally in our atmosphere and now known to be a by-product of culturing almost all aerobic (oxygen-using) organisms grown in a rich organic medium with oxygen present. However, most aerobic organisms are known today to have special enzymes. One discovered by Schonbein is called catalase, which is secreted to decompose hydrogen peroxide into water and oxygen before peroxide damages the organisms. Another enzyme, peroxidase, reacts with hydrogen peroxide and an elemental substance such as iron or sulfur, yielding water and the oxidized chemical. The nature of these two enzymes, peroxidase and catalase, remained obscure until the early 1930s. Their discovery would later have far-reaching implications when some Viking scientists postulated that hydrogen peroxide formed in the Martian atmosphere is produced as it is in the Earth's and then settles to the planet's surface, destroying any organic material that may be there. As we will see in Chapters Four and Five, no evidence of hydrogen peroxide has yet been found on Mars, but if it is, it may be crucial to understanding the early evolution of life on Earth, particularly if Martian microorganisms utilize catalase and peroxidase enzymes. It could be an ancient mechanism for organisms to have a source of water and oxygen in a world where there is little.

Another discovery made by the French chemist R. Latarjet in 1952 was that certain bacteria rendered "sterile" by large doses of ultraviolet light are restored to activity by peroxidase (Latarjet, 1952). Since Mars has a limited abundance of ozone to absorb strong ultraviolet radiation, maybe Martian organisms evolved ingenious ways to protect themselves, perhaps utilizing the peroxidase enzyme. Or as an alternative solution, if hydrogen peroxide exists in the soil of Mars, microorganisms there might also use it as a source of water if they secrete peroxidase and catalase. Without the knowledge of such organisms on Earth, this concept is not likely to persuade the many scientists attempting to explain Viking who view hydrogen peroxide only as destructive of living systems.

The true nature of atmospheric hydrogen peroxide and the role it plays in the biosphere of Earth was not yet understood at the time of the Viking Mission to Mars in 1976. Not until the mid-1980s was hydrogen peroxide studied on our planet *in situ*. These investigations provided researchers in the fields of meteorology and microbiology a new look at the way peroxide fits in with the overall scheme of life. I will return to this important subject again in Chapters Four, Five, and Eight.

In 1887, a Russian scientist working at various European universities and later at the Pasteur Institute in Paris discovered bacteria in polluted or stagnant ponds that could oxidize hydrogen sulfide—an extremely poisonous and toxic substance—breaking it down to elemental sulfur through the course of their metabolism. The significance of this discovery by Sergei Winogradsky (1856–1953) gained him the title "the founder of soil microbiology." These "green" and "purple" sulfur bacteria were found to carry on photosynthesis similar to ordinary green plants, but by using a photosynthetic pigment called bacteriochlorophyll rather than chlorophyll. However, these "chemophototrophic bacteria" do not release oxygen as a by-product of their respiration, and they use carbon dioxide as their only source of carbon.

Since the atmosphere of Mars contains about 95 percent carbon dioxide, there would be plenty of carbon for these types of organisms to thrive on. In fact, Mars has more atmospheric carbon dioxide (7–10 millibars) than the Earth (0.35 millibars). It is important to remember that atmospheric carbon dioxide is the ultimate carbon source of all living matter on Earth, and today 90 percent of the carbon dioxide produced is formed by bacterial action. Organic deposits that build up in Earth soils are constantly being degraded by microbial activity and converted into carbon dioxide. Another way certain microorganisms contribute carbon dioxide to the atmosphere is by releasing the enzyme decarboxylase—which liberates carbon dioxide from amino and fatty acids in soil. It has been calculated that all the carbon dioxide of Earth passes through the photosynthetic process, by all types of photosynthetic microorganisms and plants, in about 350 years. If all the microbes were to suddenly cease producing carbon dioxide, then within only three months all life on Earth would perish (Rabinowitch, 1951, *Photosynthesis and Related Processes*, Interscience, New York).

A recent example of just how important a role microorganisms have in the context of affecting carbon dioxide content in an atmosphere comes from the studies made inside Biosphere II—a 3.15-acre closed terrarium located in Oracle, Arizona, owned and operated by Space Biospheres Ventures, Inc. The Biosphere II enclosure contains air, water, soil, animals, and plants designed with the purpose of studying changes within jungle, desert, ocean, savanna, and salt marsh environments. Another research objective of this unusual project is to examine the role that closed ecosystems could play in the colonization of Mars. Eight researchers would spend two years gathering information on changes inside.

A report from a Biosphere II committee on biodiversity to the National Science Foundation in 1993 showed that after only 1.4 years of habitation by the eight human "Biospherians," oxygen levels inside declined from a normal of 21 percent to 14 percent. To the human occupants of Biosphere 2 this change was the equivalent of the low level of oxygen found at elevations of 17,500 feet above sea level. What happened to it? Bacteria and fungi in the fertile soil of Biosphere II were consuming the atmospheric oxygen and rapidly converting it to carbon dioxide. This situation became so critical that the eight Biospherians inside had to abort their two-year mission goal. It was later discovered that if not for the large amounts of structural concrete inside Biosphere II, the carbon dioxide levels could have been lethal. The concrete apparently absorbed much of the atmospheric carbon dioxide and produced calcium carbonate from it. While the aborted Biosphere II mission was considered less than a success, it did prove the ability of microorganisms to radically alter their environment and the biosphere, whether on Earth or Mars.

The importance of the chemophototrophic bacteria is that they are thought to be some of the most ancient organisms on the Earth. The granules of elemental sulfur metabolically deposited outside their cells become large deposits of soluble sulfur of commercial value. Some estimates from the small Cyrenaican lake *Ayn-ez-Zauya* indicate an annual sulfur production of more than one hundred tons. This translates into 1.7 milligrams of biologically produced sulfur per liter of water!

A discovery by the Viking inorganic analysis experiment (XRFS) on Mars in 1976 shed new light on its surface chemistry. The surface samples analyzed by this experiment at the *Viking 1* and *2* landing sites had clay-dust and salts along with ice in a crumbly layer of soil called "duricrust," shown to contain 20 to 50 percent more sulfur than is found in Earth soils. The thin layer of dust on this duricrust had one hundred times the sulfur content of the Earth (Clark, B. C., 1977). Understanding the nature of duricrust on our planet may hold the key to understanding any widespread biota on Mars. Terrestrial duricrust, or caliche as it is sometimes called, is formed by a complex process involving both geological and biological contributions to desert soils, as we will explore in this book. Duricrust on Mars was found by both the *Viking 1* and *2 Landers*—each separated by four thousand miles. It seems reasonable to assume that duricrust is a planet-wide process on Mars.

Almost all of the phototrophic bacteria prefer a moist environment such as ponds, rivers, and lakes, along with moist rocks and sand, so how could they develop and thrive on a planet like Mars, which according to many scientists today, has a more arid environment than the harshest deserts of Earth? Not until 1969, when *Mariner 9* conducted its orbital survey of Mars, was evidence found that ancient river beds, lakes, and even a shallow ocean existed in Mars' distant past. Later, evidence would be found to support the notion that Mars is capable of sustaining liquid water on its surface today when its atmospheric pressure is from 6.1 to 10 millibars and temperatures above freezing. The atmospheric pressure at the *Viking 1* landing site varied between 6.8 and 9.0 millibars with temperatures of –14 degrees Celsius (+7 degrees Fahrenheit) at midday in the summer and a predawn temperature of –77 degrees Celsius (–107 Fahrenheit). The *Viking 2* site had atmospheric pressures between 7.3 and 10.8 millibars.

Scientists call the conjunction of 6.1 millibars total atmospheric pressure and 0 degrees Celsius the "triple point of water," at which the liquid, solid, and vapor states of water coexist. However, temperatures on the surface of Martian rocks and soil can reach 36.6 degrees Celsius (98 degrees Fahrenheit) in the summer equatorial regions of Mars (T. Z. Martin, 1996, personal communication) and remain that way for hours per day and weeks on end, as measured by the *Viking Orbiter* instrument called the Infrared Thermal Mapper. This instrument was designed to make thermal inertia (temperature) measurements of the Martian surface from orbit.

The sulfur bacteria are organisms known to display bright colors such as red, purple, green, orange, and even brown because they contain a great variety of carotenoid pigments. These pigments function within the cell as accessory light harvesters and also serve to protect the bacteriachlorophyll (which reacts the same way chloro-phylls do in green plants) from photo-oxidation by the ultraviolet light. Perhaps their most startling characteristic (with special refer-ence to the species known as gliding green bacteria) is they can thrive both anaerobically (without oxygen) in the light, and aerobically (with oxygen) in the dark! If organisms such as the sulfur bacteria evolved to withstand Martian temperature variations and developed mecha-nisms for storing water during times when it is available, then their presence could account for a number of color variations that exist in the soil and rock coatings (five soil colors) during Martian spring and summer. One species of this bacteria on Earth, Chloroflexus, can be green, brown, and even orange while growing in extremely bright light under water, but is poisoned by exposure to oxygen in our planet's atmosphere when removed from its environment.

Green and purple sulfur bacteria have been intensely studied for decades. Recent investigations have demonstrated the ability of these organisms to survive in low temperatures of 8 to 10 degrees Celsius and lower. They also have been known to inhabit hot sulfur springs having temperatures of 60 to 80 degrees Celsius and greater. The sulfur bacteria are found in a variety of environments now, includ-ing moist sand and silty clay soils, and covering rocks and stones under a layer of algae. The ability of purple sulfur bacteria to grow under a layer of algae or ice is another fascinating aspect of these microorganisms. While green plants in their photosynthesis utilize light rays in the range of 700 nanometers, the purple bacteria con-duct photosynthesis by absorbing light rays in the 900-nanometer range.

If cousins of these organisms exist on Mars it could explain the vast range of colors there. Large areas of purple are often seen in the *Viking Orbiter* images newly reprocessed by the U.S. Geological Sur-vey's Astrogeology Branch in Flagstaff, Arizona. Conceivably, a sea of living organisms similar to algae and purple bacteria might be in the upper soil layers of Mars. As the seasons change from summer to fall, the green algae give way to a colder temperature-tolerating pur-ple sulfur bacteria, changing the colors of the landscape of Mars in the process from a greenish to a more purple hue. Evidence of color changes on the surface of Mars from the *Viking Landers* has also been observed and documented, as I will cover extensively in Chapter Five. Will the extremely sensitive color camera on the *Mars Pathfinder Lander* spacecraft demonstrate beyond doubt the color changes of Mars when it lands in July of 1997?

Another discovery made by Winogradsky was that the energy derived from the oxidation of ferrous iron could serve as the sole energy source for growth of a group of microorganisms that have come to be classified as chemoautotrophs. This discovery was fantastic—an organism that could use iron as an oxidant while growing on elemental sulfur. Since this sulfur was discovered to build up on the surface of sulfide minerals during their oxidation by oxygen, the unique ability of bacteria to oxidize elemental sulfur using ferric iron as an oxidant was of enormous interest to geochemists looking for new ways of leaching sulfide ores and developing acid mine drainage systems.

A chemoautotroph is defined as a bacterium that has the capacity to derive its energy by oxidizing inorganic materials such as minerals. Thus chemo = chemical. An autotroph is a bacterium that uses carbon dioxide as its sole source of carbon. Having these two unique capabilities combined into one microorganism demonstrates the versatility and uncanny ability of life to organize and evolve with changing environments. A geochemical result of the activity of the chemoautotrophic bacteria is the synthesis of organic material from inorganic compounds in the absence of sunlight or photosynthesis. The relevance that this would have to exobiologists today and Mars is that in 1976 the Viking inorganic experiment (XRFS) found evidence suggesting that the soil of Mars is rich in iron, ferric oxides, maghemite, and perhaps carbonates such as calcite and sulfate minerals. However, the oxidation state of the abundant sulfur on Mars is not known from the XRFS data, but if peroxides are on Mars, then the sulfur would be oxidized.

This is the same composition of materials that existed on the ancient Earth when life first arose or implanted. Unfortunately, this Viking instrument did not allow for direct determination of minerology on Mars; instead these compositions were arrived at through modeling techniques considered by some to be nothing more than educated guesstimation compared to the analysis made on the returned Apollo lunar samples or meteorites. However, the estimates were enough to indicate that Mars has the inorganic materials that chemoautotrophic organisms could use to build cells.

The *Viking Lander* camera system was superior to the two cameras on board the *Viking Orbiters* because the Orbiter cameras were designed for engineering support of the Landers in search of a suitable landing site with a very limited fifty-meter-per-pixel resolution. Identification of Martian minerals from orbit would be unlikely because of the camera limitations. Attempts were made by the imaging team with the Lander cameras to determine the relative spectra of the rock materials using various filter diodes that included the far-infrared. The primary results of this study—though the Viking cameras were plagued by color calibration problems—indicated various amounts of strange "weathering coatings" on the rocks, along with the suspected spectrum of various iron-oxide forms. It was discovered that in addition to the ordinary iron oxides like hemetite, other iron oxides such as geothite—a hydrated iron oxide containing molecules of water in its crystal structure—also exist, along with evidence of magnetite, a mineral sometimes produced by bacteria. Chemoautotrophic organisms would find Martian iron oxides very palatable!

A species of *Thibacillus acidophilus* isolated in 1972 showed the ability to grow "autotrophically" on elemental sulfur in addition to "heterotrophically" on a variety of organic compounds including sucrose, fructose, ribose, glucose, maltose, xylose, fumaric and succinic acids, mannitol, and even ethanol! The significance of these organisms to Mars and life would not be appreciated until twenty years after the Viking Labeled Release experiment looked for life on Mars. Many of the scientists who were critics of this experiment said that it could only detect heterotrophic organisms and had failed to detect organisms that might be able to use both inorganic and organic compounds for growth. The Viking Labeled Release experiments utilized simple amino acids and simple carbohydrates, e.g., glycine, L&D alanine, L&D lactate, formate, and glycolate, all of which are produced in the Miller-Urey reaction and hence were present on early Earth and probably early Mars (if the reducing atmosphere theory is true). These simple compounds are incorporated and used by evolving life forms and they persist in present biochemistry.

A. I. Oparin himself in 1938 postulated that the earliest organisms were fermenters, organisms that used enzymes to extract energy from carbohydrates for energy maintenance. The prebiotic environment of Oparin's Ocean Scenario or "primeval soup" would have contained Miller-Urey products including sugars. Evidence of aldehydes would be found only in the dark areas of Mars by Dr. William M. Sinton of the Harvard and later Lowell Observatories using infrared detection techniques in 1956 and 1960. If they are indeed present on Mars, aldehydes and formaldehyde could be possible sources for simple sugars. (Sinton's work is covered below in this chapter.)

Since the discovery of chemoautotrophic organisms and others even more versatile, it is ascertained that the interior of our planet is not sterile. Microorganisms are consuming, making, altering, and redistributing substances in the hydrosphere, atmosphere, and lithosphere of our planet. Weathering of ore deposits, both in iron and sulfide ores, has been occurring since antiquity. Penetration of microorganisms through cracks and crevices of our planet has now been confirmed at depths miles below the surface of the lithosphere—this invasion has been occurring since the first life-forms evolved. Of course, there is the distinct possibility that life may have thrived at some depth in the lithosphere of our world before appearing on the surface. On the other hand, life may have begun or been transported in such icy dust-strewn solar system objects as comets. Mars and the Earth may have undergone a cycle of organisms submerging into their prospective lithospheres and being alternately pushed to the surface again by hydrothermal activity, volcanism, seismic activity, and outgassing of volatiles.

Perhaps the most startling find regarding Winogradsky's sulfur-iron oxidizing bacteria is that there is a widespread occurrence of barite (barium sulfate), a common gangue mineral in medium- and low-temperature hydrothermal veins in Archean sediments of the Isua area in West Greenland. This suggests that sulfate was in the

Earth's oceans 3.5 billion years ago, now thought to be the result of photosynthetic sulfur bacteria. Also, rocks isotopically dated from this region that are 3.8 billion years old demonstrate carbon fixation, telling us that microorganisms were metabolizing during a time in solar system history known as the late bombardment period when the Earth, Moon, and planets were being impacted frequently by giant comets, asteroids, and left-over debris from the solar nebula. In fact, this is a time period when the planets themselves are thought to have been still solidifying from a molten mass (Schidlowski, M., 1983). It has now been paradoxically postulated that life on Earth may be older than the oldest known rocks (Pflug, H. D., 1986).

Further evidence of this stunning discovery would be announced by Dr. Gustaf Arrhenius of the Scripps Institution of Oceanography at San Diego in November of 1996 and results published in the November issue of *Nature*. Dr. Arrhenius is the grandson of Svante Arrhenius, whose work in the early nineteenth century led him to believe that life is implanted on the planets from elsewhere in the cosmos by meteorites and interstellar dust—the theory of Panspermia. Dr. Gustaf Arrhenius has come one step closer to proving his famous grandfather right about Panspermia—his discovery of 3.85-billion-year-old rocks containing evidence for microbial life from the Isua area of West Greenland. This pushes the date back for life on Earth another four hundred million years beyond any previous study. Questions arising from this discovery are: how did life develop so quickly on a planet still cooling from a molten phase? How could this life continue to evolve in a world of constant heavy bombardment by comets, meteorites, and asteroids? Was life falling in on the planets from space as Svante Arrhenius suggested in 1907?

The Search Begins

In 1907, the famous Swedish chemist Svante Arrhenius (1859–1927) wrote in his book *Das Werden der Welten* about his theory of how life is transferred from one planet to another by means of pressure from light rays, borrowing some of his concepts from Richter and Helmholz years earlier. Arrhenius believed it was possible for living "germs" combined with interstellar and interplanetary dust grains to seed planets throughout the galaxy. His critics maintained that it was highly unlikely that such a weak force as a light ray would be capable of overcoming a planet's gravity. Furthermore, they thought, life could not survive the intense radiations of space, especially those of the short-wave ultraviolet variety, which during Arrhenius' time were thought to be lethal to all living things.

Arrhenius postulated in 1909 that air currents could take volcanic and other dust to the upper layers of the atmosphere over one hundred kilometers in altitude, where they could be ejected into space by electrical discharges in Earth's atmosphere. From here the particles would be propelled by the force of the "solar wind," an invisible wind of superheated solar gas that flows continuously out of the Sun. If this process were relevant to Earth dust and bacterial spores, he thought, then it was conceivable that it could happen throughout the galaxy on planets that had developed life. He even calculated the speed that spores on dust grains could travel on solar wind if they had a maximum diameter of 150 to 200 millimicrons—a distance spanning the orbit of Pluto in fourteen months! What Arrhenius could not have known is that just such a powerful electrical discharge mechanism has recently been found on Earth accidentally by an experiment called the Burst and Transient Source (BATSE) aboard NASA's Compton Gamma Ray Observatory. The experiment was designed to look for gamma-ray bursts in the interstellar regions of space.

Since 1993 scientists using the BATSE have recorded thousands of gamma-ray bursts coming from the cloud tops of thunderstorms. Space Shuttle astronauts have reported seeing strange flashes of light emanating upwards from thunderstorms. In fact, three different types of these objects are now known—"Red Sprites," which can reach altitudes of one hundred kilometers (sixty miles), "Elves," doughnut-like bursts lasting only eighty-five milliseconds or less and occurring with sprites, and rarest of all, "Blue Jets," which can travel at speeds of one hundred kilometers per second! The energy of these new forces is up to thirty times more powerful than the surrounding lightning. Arrhenius had again brought the concept of Panspermia to the forefront of the question "from where did life come?" Arrhenius did not propose, however, how life began. Such a concept would shortly be provided by another, A. I. Oparin, in 1925 with his small Russian booklet simply entitled *The Origin of Life*, which was revised and expanded in 1936 to a full-length book of the same title and sold internationally. (Oparin's hypothesis is covered in my Introduction.)

Arrhenius was a brilliant physical chemist who won the Nobel Prize for chemistry in 1903 for his electrostatic dissociation theory. He went on to become the director of the Nobel Institute for Physical Chemistry in Stockholm, Sweden. Caught up in the Lowell-Mars canal controversy, Arrhenius set out to demonstrate that the Mars phenomena of canals and the "wave of darkening" were anomalies unrelated to life. His idea centered on chemical changes in the soil of the dark areas of Mars that were "activated" by water from the melting polar ice cap and surface minerals such as water-absorbing hydroscopic salts. Arrhenius felt this explanation would account for the seasonal color changes as well as the appearance of "canals." It is interesting to note that even today, geologists, chemists, and biologists all tend to explain observed phenomena according to their particular expertise in science, especially when it comes to the subject of life on another planet!

Arrhenius further infuriated Percival Lowell when he described the canals as nothing more than volcanic cracks and fissures. Was Arrhenius successful at swaying those who continued to insist that life might be responsible for the seasonal color changes? Perhaps for some, but most astronomers of the time were too intrigued with the possibility of "life on another world" to enthusiastically embrace his theories, and the search for what could cause the seasonal "wave of darkening" would continue with Lowell himself leading the way until his death in 1916.

Percival Lowell was a determined man, and if Mars had canals and vegetation, he was going to prove it. A series of observations by three key Lowell Observatory staff members again pushed Lowell's concepts to the forefront of innovation and scientific controversy. The first big push forward came with Lowell's hiring of Carl Otto Lampland in 1903, who was a pioneering planetary photographer (and in fact won a medal from the Royal Photographic Society in 1907 honoring his photographs of Mars). In 1905 Percival Lowell proclaimed that Carl Lampland had succeeded in photographing the elusive canals of Mars. As Lowell went from publisher to publisher trying to get these photographs printed, none of the editors who reviewed them could see the canals on the photographs. There were hints of something linear on them but nothing convincing. Therefore, Lowell made a series of drawings at the same time Lampland made his photographs of Mars to demonstrate what could be seen visually as compared to photographically. Although Lampland's photographs were extremely good and showed the large darker features of Mars, Lowell's claim that he had photographed "definitely" the canals of Mars became another joke among professional astronomers opposed to his theories. Just another desperate attempt to try and sway his peers, they thought.

Vesto Melvin Slipher was hired by the Lowell Observatory in 1901 for his experience in astronomical spectroscopy. Percival Lowell instructed Slipher to begin a spectroscopic search for chlorophyll from 1905 to 1907. The search ended with the conclusion that the light coming from Mars was too weak to be analyzed by standard astronomical spectroscopy techniques. However, no one yet realized that the dawn of a new era had begun—the application of remote sensing techniques in the search for life on other planets.

Vesto Melvin Slipher's younger brother, Earl Carl Slipher, was hired by Lowell in 1906 under the graduate Lawrence Fellowship program that permitted graduate students from Indiana State University's astronomy department to work in a professional observatory environment. E. C. Slipher, as he came to be known, took a serious interest in planetary photography while at Lowell Observatory, and during the Mars oppositions of 1924 and 1926, he took a series of photographs that depicted the seasonal changes on Mars, or the so-called "wave of darkening." Unlike Otto Lampland's photography and search for the Martian canals, E. C. Slipher's pictures clearly showed the dark areas of Mars growing in size as the polar ice cap melted and disappeared. Unfortunately, Lowell had died by this time. In E. C.'s set of Mars photographs, astronomers could study the wave in detail progressing from Mare Erythraeum to Aurorae Sinus, then to Margaritifer Sinus and Meridiani Sinus. It was also observed in these photographs that the areas on Mars that had darkened early lightened as the season progressed. The wave of darkening was found to differ from the way terrestrial vegetation proceeds seasonally. On Earth, when winter turns to spring, vegetation first responds to the warming in the mid-latitudes, and then the response progresses toward the poles with the rising temperature. On Mars, it appeared that the dark regions were responding to transport of moisture from the polar regions toward the mid-latitudes—opposite the trend on Earth.

In 1924, Vesto M. Slipher looked a second time for evidence of chlorophyll on Mars, this time utilizing the technique of photography combined with infrared color screens. These color screens were an early version of today's colored astronomical filters, which permit planetary observers a means of filtering out unwanted wavelengths of light and thereby allowing details otherwise obscure or reduced in contrast to stand out boldly. It was then discovered that if infrared color screens were used to photograph vegetation on terrestrial landscape pictures, that vegetation appeared extremely bright against other objects such as mountains, due to the absorption of chlorophyll in the blue and red parts of the spectrum.

Slipher applied this knowledge to photographs of Mars. He first photographed Mars using a red color screen, which rendered the dark regions of Mars even darker. In fact, astronomers today still use red "filters" to study the dark areas of Mars. However, in photographs taken with infrared color screens, if indeed the dark maria of Mars were due to chlorophyll, any vegetation would appear very bright against the background. Since Vesto M. Slipher's photographs did not detect this infrared brightening, the idea that vegetation was causing the wave of darkening became doubtful.

The seasonal appearance of the Martian wave of darkening was relentless. By the mid-1930s, Earl C. Slipher had mastered black and white photography of Mars. Although his photographs did little to indicate the canals as a real photographic anomaly, his thousands of pictures did show quite clearly the wave of darkening as it progressed through the Martian seasons. Mars was by no means in danger of being ignored by astronomers in the 1930s, and a few surprises would still be discovered before the decade was out.

Many scientists studying Mars today believe that it is a volcanically inactive planet, but on June 4, 1937, a Japanese astronomer by the name of Sizuo Mayeda was observing Mars in the region of Tithonius Lacus. This region is situated near the Martian equator, as is a huge rift valley system more than sixty miles wide and four thousand miles long. Not far from this great rift valley are the Tharsis volcanoes, Arisa Mons, Pavonis Mons, and Ascraeus Mons. While studying Mars, Mayeda suddenly witnessed an amazing event—a brilliant flash at first, and then a flickering of light for a period of five minutes. Even closer to Mayeda's point of observation of the explosive event is the Martian volcano, Tharsis Tholus. Had this Japanese astronomer been the first to witness an eruption on Mars?

In the summer of 1939, Edison Pettit of the Mount Wilson Observatory was looking at the canals of Mars with a 60-inch telescope. Keep in mind that the canals were still a very real problem for astronomers of the time period, and the more they were scrutinized by astronomers, the more mysterious and elusive these "canals" became. Schiaparelli and Lowell never mentioned any unusual colors associated with the canals. Instead, to them the canals always appeared dark or gray. On that summer night in 1939, however, Edison Pettit made a most fascinating discovery. The thin lines crisscrossing Mars seemed to be olive green in color. Unfortunately, even though Pettit took some color photographs, the images of Mars were so small on film (three millimeters) that the canals so apparent to his eye were not on the film, just as in the case of Otto Lampland's photographs of Mars in 1909. These subtle features just could not be recorded on film, it seemed.

In another part of the world, Gavriil A. Tikhov, a prominent astronomer at the Pulkovo Observatory in Russia, was probably the first true astro-biologist to apply the sciences of astronomy and botany together in a systematic search for the types of organisms that could exist on other planets, with Mars being his life-long focus. He was the first astronomer to compare spectroscopically the Martian surface features and the various spectra of terrestrial flora. He called this new branch of astronomy "astrobotanics." In 1947, Tikhov formed the Sector of Astrobotany at the Institute of Physics and Astronomy of the Academy of Sciences in Kazakhstan. The concept of this elite group of astrobotanists was to study the behavior of vegetation in the extreme areas of Earth such as the Arctic tundra and high mountain areas, and apply those studies to the question of whether or not certain species might be able to survive on Mars.

Tikhov was annoyed with conventional "geocentrism in biology." He stated, "by it we mean that Earth is a kind of standard body most favorable for life, to some extent central, and deviations from its physical properties in either direction render now impossible the origination and existence of life ... geocentrism in astronomy refers to the theory stating the Earth is the center of the entire visible universe" (NASA TT F-819, 1975). Tikhov was outspoken in his criticism of astronomers who posed the possibility of life "somewhere" in the depths of the universe, but could not see how it might also have begun on Mars. Of these astronomers he wrote,

> Opponents of the idea that even plant life exists on Mars do not doubt that somewhere in boundless space, life does exist, but we cannot now either show this by observation, nor in any foreseeable time, and therefore they hope that if the existence of life on Mars will be proven, this will only be a special case lacking general, fundamental importance. In this way, biological centrism will be saved. How similar is this to all the attempts to save geocentrism after the remarkable investigations of Copernicus, Kepler, and other leading figures in astronomy. (NASA TT F-819, 1975)

Though he spoke these words decades ago, Tikhov could not possibly realize that some scientists who actually had experiments touch down on the surface of Mars in 1976 would still defend the "biocentrism" of life on Earth, even with evidence to the contrary in hand, as we will cover in detail later.

Another fascinating aspect of this astronomer from Russia was his ability to think around scientific obstacles preventing the possibilities for life on Mars. For example, we have discussed already the attempts of Lowell, Slipher, and other American astronomers searching for chlorophyll on Mars. Tikhov himself looked for chlorophyll on Mars in 1918–1920 using a 15-inch refractor and an infrared color screen. He calmly stated that he had no technical difficulties in his search but added, "I did not find chlorophyll"!

For many astronomers, not finding evidence for chlorophyll meant no vegetation on Mars. For Tikhov and his team of "astrobotanists," their solution was to find another way around the problem and illuminate a new concept. Tikhov deduced that since it was much colder on Mars than the Earth, that life of a terrestrial plant in a temperate climate would absorb a small region of red rays corresponding to a chlorophyll absorption band in the infrared. In the severe climate of Mars, he thought, this would be insufficient:

> A plant on Mars must absorb, with the aid of special pigments, the heat rays of the sun which adjoin on one or the other side of the main chlorophyll absorption band, and thus, this band becomes only faintly detectable. (NASA TT F-819, 1975)

G. A. Tikhov set out to prove his theory by checking the spectra of vegetation growing in severe-climate regions of the subarctic. From his studies of these harsh environments, Tikhov pointed out that many areas on Mars observed through the telescope are blue in color. This fact has been confirmed by *Viking Orbiter* images reprocessed in the 1990s by the United States Geological Survey's Astrogeology Branch, in which gigantic regions of blue, blue-green, and purple soil meet with yellow-orange deserts.

Tikhov's study of Earth vegetation in harsh climates led him to many species of vegetation that had blue pigments, such as the blue poppies in the Himalaya and blue spruce trees of Canada. He was on the right track. He said of his discovery:

> If a plant strongly absorbs red and contingent orange, yellow, and green rays which in fact represent one-third of solar heat, then in the plant-reflected "cold" light consisting of deep blue, blue, and violet rays—then blue, deep blue, and even violet light plays a major role, and the plant obtains the corresponding coloration. That is why the Martian dark regions are blue and dark blue. (NASA TT F-819, 1975)

Tikhov went on to write two books (1949) entitled *Astrobiology* and *Astrobotany* which detail his theories on the paleobotanics of Mars and the adaptive properties of plants. The books also considered such concepts as organisms that were able to develop as a by-product of metabolism an antifreeze such as glycerol. The fresh ideas and concepts in Tikhov's books caught the attention of both biologists and astronomers worldwide and were very influential for the development of the cutting-edge science we know today as exobiology.

THE NEXT WAVE OF MARTIAN discovery would require the special features of a highly refined group of astronomical instruments which are attached to telescopes. These include the monochromator, a narrow-waveband filter used in combination with a telescope to provide information on one limited region in the electromagnetic spectrum. The monochromator can be used to detect the faint spectrographic signatures of organic materials. Another of these instruments is the polarimeter, a device used to measure the angle of refraction of light waves vibrating in one particular plane, known as the plane of polarization. A polarimeter can be used effectively to study planets, allowing astronomers to measure some properties of the surface material down to microscopic features. This incredible device can also show if a planet has a smooth surface; has a surface formed by distinct grains; consists of coarse or polished grains; is absorbent to a certain degree; and if it has a powdery surface with absorbing grains.

Jean-Henri Focas, a Greco-French astronomer, took part in one of the most detailed polarization studies of Mars ever attempted. Using the 25-inch refractor of the Pic du Midi Observatory in the Pyrenees mountains of France, he along with his colleagues made more than four thousand measurements using this technique, beginning in 1948. Focas used magnifications of up to 900 power on the 25-inch Pic du Midi refractor, which enabled him to get very accurate data from different regions on the Martian surface. The study showed interesting variations in polarization. First, not much difference was noted between the polarization step from the dark regions of Mars and the bright orange regions. Then it was found that the polarization varied with the Martian seasons! A distinct and unmistakable difference was noticed in the spring northern hemisphere, then summer, and with a constant polarization shown in the winter, but a fluctuating polarization when spring arrived. The French astronomers noted that the same process exists for the southern hemisphere of Mars. What could be causing such a peculiar effect? they asked. They came to the astounding conclusion that whatever was causing these seasonal variations on the surface of Mars was on the microscopic scale.

Another thought came to them. Physical or chemical changes in the soil could not explain these bizarre changes, and in effect, the theory that Svante Arrhenius had regarding a chemical change in the soil as a result of minerals reacting with moisture release from the polar ice caps would have to be reconsidered. For the first time, it was postulated that organisms on a microscopic scale—or as they were called then, "animated organisms"—might cover the entire surface of Mars and be responsible for the planet-wide seasonal color changes.

This work at the French observatory would be directly responsible for astronomers starting to ask questions regarding microbiology. In fact, these astronomers began to look at the spectra of hundreds of possible candidate bacteria they thought might be good terrestrial analogs for life on Mars. They looked at bacteria on Earth called chromogens, which are organisms that can produce vivid colors, and then matched their polarization data. Perhaps the most significant organism they found was a species of algae that covers ice, snow fields, and mountains, known as *Chlamydomonas nivalis* (nivalis = snow), or snow and ice algae. This organism is resistant to intense ultraviolet light and extreme cold, and has brilliant colors such as green, orange, red, and even blue. *(See Color Plate 2.)*

The relevance of the snow algae discovery would only be fully appreciated twenty years after *Viking* set down on the surface of Mars. Dr. Ronald W. Hoham, a biologist from Colgate University of Hamilton, New York, has studied these interesting organisms since the 1970s and has written a book to be released in the spring of 1997 called *Microbial Ecology of Snow and Fresh-Water Ice with Emphasis on Snow Algae* (University of Arizona Press), together with another microbiologist, Dr. Brian Duval of the University of Massachusetts. In the book, Hoham suggests that "true" snow-ice algae have an optimal growth range at temperatures below 10 degrees Celsius. The scientist further attested to the resilience of these hardy organisms, stating that materials such as salt and dust might be responsible for depressing the freezing point, and that salt may be a mechanism to prevent evaporation in a low-humidity environment.

Other factors making these organisms good Martian candidate models include their habitat around tree well basins (depressions that form around tree trunks) and similar depressions around rocks called "suncups." Many of the rocks around both Viking Landers appear to have "suncups" around them (called "moats" by Viking-era scientists). Suncups on Earth are formed by both physical and biological processes.

Snow-ice algae are often found concentrated in horizontal ice layers several centimeters below the ice and snowpack surfaces. Other information on snow-ice algae is that they depend on dust and debris as important sources of nutrients while living in ice and snow. *(See Color Plate 3.)* Also, *Chlamydomonas nivalis* was observed to accumulate heavy metals in its cells thousands of times their concentration in surrounding snow.

Just what does this have to do with the Viking discoveries? Dr. Henry J. Moore of the U.S. Geological Survey was the Principal Investigator for the *Viking Lander* Physical Properties experiment. Moore and his team used the Viking surface sampler (the same one used to gather biology samples on Mars) to examine the texture of the soils on Mars. What they found was very interesting. Although it appeared many pebbles were present at the *Viking Lander 1* and *2* sites located on Chryse Planitia and Utopia Planitia, when Moore's team tried to pick some of these up for analysis, they kept breaking apart. In other words, no pebbles were found. To make the situation even more interesting, when soils were sampled for particle size measurements, it came as a surprise that only silt-sized particles could be found, not sand or larger grains (Moore, H. J., 1977). The soil just below the surface was, as Moore put it, "smaller lumps that are weakly cohesive clods of material with grains finer than 0.2 cm." The clods became known to Viking scientists as "duricrust," another desert

RICK KLEIN/CORNELL UNIVERSITY

Viking Lander *image showing "false pebbles," which crumbled when picked up by the sampler arm. Also, in the center of the photo, an example of duricrust can be seen winding through the soil.*

environment formation found in most terrestrial deserts (Moore, *et al.*, 1977). The Martian duricrust, found at both Viking Lander sites, has a high sulfur content, probably as sulfate. Chlorine is also present, probably as chloride salts (NaCl). The upwelling of moisture through the lithosphere of Mars mixes these salts with iron-rich clays. This clay soil acts as a sponge, storing large quantities of water.

Perhaps one of the most interesting discoveries about the composition of duricrust on Mars is that bromine was detected at a few hundred parts per million and is suspected of being more abundant in deeper layers of the Martian crust. On Earth, bromine compounds photochemically react with surface ozone in the Arctic during spring, destroying large quantities of ozone. It has been discovered that Arctic ice-algae emit significant quantities of biogenic bromoform, which gets photochemically converted to organic bromine. Globally significant amounts of organic bromine compounds are now believed to come from Arctic ice microalgae (Sturges, W. T., 1992). Biogenic bromoform emissions are maximal in winter and minimal in summer. A species of brown Arctic seaweed called *Ascophyllum nodosum* also, through enzymatic activity, releases large amounts of organic bromoform *(Environmental Science and Technology*, Vol. 25, No. 3, 1991).

The Martian duricrust represents the upper one inch of soil. On Earth, true duricrust, sometimes called desert crust or caliche, is associated with microorganisms and the capillary rise of liquid water which carries salts with it, forming the duricrust layer. Microorganisms such as blue-green algae, autotrophic and heterotrophic bacteria, fungi, and actinomycetae found in terrestrial desert crusts also contribute carbonates through biochemical precipitation of calcium carbonates in desert soils. Evidence of this type of biomineralization has been documented since 1968! (Krumbein, 1968)

The discovery by the *Viking Lander* x-ray fluorescence spectrometer (also called inorganic chemical analysis instrument) of chlorine, bromine, and iodine as halogen salts in the soil of Mars should have been cause for Viking scientists to consider a biogenic origin. Chlorine, bromine, and iodine appear as gaseous compounds in Earth's atmosphere, such as HCl, ClO, HBr, and BrO. They also appear as halogen-bearing organic compounds in marine aerosols and rainwater. However, the dominant source of these halogens is the oceans. The source of chlorine in our atmosphere is NaCl from sea spray (Blanchard and Woodcock, 1957). Iodine comes from the release of iodine gas in iodine-rich organic material of sea-surface films ejected in sea-spray (Miyake and Tsunogai, 1963; MacIntyre, 1974). Bromine, chlorine, and iodine then get deposited on soil surfaces through precipitation as rain, snow, dew, or fog, where they build up as salts over time periods of hundreds of thousands of years. Halogen salts in the soil of Mars could be interpreted as evidence for an ocean that existed in Mars' distant past. The organic origin of halogen salts has been known since the 1960s, so why was NASA relying so heavily on the opinions of chemists and geologists who were reluctant to consider biological interpretations of the observed phenomenon?

Moore made another keen observation regarding the Martian rocks—they appeared to him to be coated with a substance known on Earth as "desert varnish" (personal communication, Moore to Levin, 1979). Desert varnish, sometimes referred to as rock varnish, is a very hard, dark, but shiny coating on rocks and pebbles which is produced by chemical and biological weathering processes after many thousands of years. Desert varnish has been found in almost all of Earth's hot and cold deserts including the Atacama in Peru—the world's driest, and Antarctica—the world's coldest. For decades geologists and chemists maintained that desert varnish was a result of only chemical and physical weathering of rocks. Evidence of a biogenic origin of desert crust and desert varnish had been known since 1963 (Schwabe, 1963) by biochemists. The outer surface of the varnish can greatly vary in appearance, from a dark red to a lustrous black. Since the Viking Mission to Mars, geomicrobiologists have now confirmed that almost all desert varnish is a result of bacteria, fungi, algae, and lichens.

When desert rocks get a slight dampening from fog or dew, spores from lichens begin to grow while pulling chemicals from the rocks under the energy of the Sun. As the lichens and chemicals slowly cover the rocks to form a thin microscopic absorbent layer, each time the rock is moistened it adds new layers. This can take centuries. Desert varnish can be found on ancient man-made structures such as the pyramids of Egypt and in the American Southwest, where petroglyphs (rock drawings) were scratched by ancient Indian peoples into the surfaces of rocks with desert varnish coating on them. The action of the lichens pulling chemicals up out of rocks and forming desert varnish is also a means by which soil is enriched, making it possible for other species of organisms to survive in it later.

An assortment of *Viking Lander* photographs depicting what appears to be desert varnish can be found in the NASA publication *The Martian Landscape* (NASA SP-425, 1978). Most of the photographs in the book are black and white, and the varnish appears on the Lander images that were taken at sunrise.

Moore goes on to say in his scientific paper that other Viking experiments—the Molecular Analysis experiment (GCMS) and the Pyrolytic Release experiment (PR)—demonstrated that the Martian soil contained "large amounts of water." His most astounding remark came under the heading "Temperature." He noted that the sample collector head (which had a temperature sensor on it) and *Viking Lander 2* footpad temperature sensor registered the temperature of the air to be compatible with liquid water. The Martian atmospheric pressure was above 6.1 millibars, making it possible for liquid water to exist. However, the temperature just below the surface of the soil being measured by the *Viking Lander 1* footpad sensor (buried 1.5 cm) was substantially higher before sunrise! (Moore, et al., 1977)

Aggregated clods of soil somehow cemented together; no sand-size particles or pebbles; water in the soil. What did it all mean? How could subsurface soil temperatures on Mars rise before dawn after the intense cold of a Martian night with a temperature of –77 degrees Celsius? For the answer to this all-important piece of the "life on Mars" jigsaw puzzle, I ask readers to wait until Chapter Nine of this book, "Life After Viking: The Evidence Mounts," guest-written by Dr. Gilbert V. Levin.

The early comparison of snow and ice algae in the 1950s to a Martian analog was a great step forward and one that seemed more plausible than the vast fields of vegetation previously envisioned by Lowell and others. In fact, the snow-ice algae, Chlamydomonas, would turn out to have an even more adaptive element to the Martian environment in its favor than anyone could have possibly realized until recently—it can scavenge, break down, and utilize hydrogen peroxide. (I will return to this important subject in the next chapter.)

The line of reasoning about algae and bacteria that French astronomers had would lead indirectly to the most exciting astronomical announcement of the next twenty years, that of American astronomer William M. Sinton's "Spectroscopic Evidence of Vegetation on Mars," published in the September 1957 issue of *The Astrophysical Journal*. Mars fever would again be elevated to a level of interest not witnessed since the time of Schiaparelli and Lowell. Leading up to Sinton's discovery were a few observations of Martian phenomena that rattled preconceived ideas of how things work on Mars.

A flurry of astronomers in 1952 witnessed one of the most amazing events ever to be seen on Mars—the gradual growing of a huge new dark area on Mars never before seen and estimated to be the size of France! *(See Color Plate 4.)* This anomalous dark area appeared in a region where there was just a light-colored desert area northeast of Syrtis Major. Mount Wilson and Palomar Observatory astronomer

Robert S. Richardson described the color of this new dark area or "maria" as "light green or gray green, no blue tint could be discerned." (Meeting of the American Astonomical Society, in New York City, December 28, 1956.) Dr. Frank B. Salisbury, professor of plant physiology at Colorado State University, stated in a lecture on possible life on Mars and the new dark maria that had formed:

> This observation was so astounding that we could even ponder the thought of a Martian intelligence able to conquer part of the "desert" for itself with the aid of agronomy. (NASA TT F-819, 1975)

Salisbury was speaking facetiously about intelligent beings on Mars but continued:

> Sometimes the Martian dark regions are covered with a layer of yellow dust, but several days later they appear again. If they consist of Martian organisms, these organisms must either grow through the dust or shake it off. I have long been amazed at the density of the Martian dark regions compared by the deserts surrounding them. If the "maria" photographed so well through a red filter, then this means they consist of organisms covering the soil in a solid layer. This is quite evident if we observe our desert regions from an aircraft at the altitude at which individual plants cannot be differentiated. (NASA TT F-819, 1975)

Salisbury made a good point here that can be applied to the orbital photography of Mars made by Russian and U.S. spacecraft. If the wave of darkening and other maria of Mars were caused by microorganisms, they would have little chance of being detected from orbit. Ninety percent of the Viking Mission orbital photographs were black and white, and it is known that the color photographs from orbit were calibrated to what the imaging team "thought" the colors of Mars should be, rather than how they registered on the calibrated Viking imaging system, as I will discuss in Chapter Three. Analysis of Viking imaging data has shown that atmospheric aerosols and dust, when present, will cause a lack of contrast in visible wavelengths (Thorpe, 1979). What does this mean exactly? It means that visible light scattering of dust and ice crystals in the Martian atmosphere severely limits what *Viking* or *Mariner* saw from orbit because of aerosols present in the atmosphere. In fact, many surface features of Mars are obscured by clouds, hazes, and mists, and these blend in so well with the landscape in the black-and-white images that sometimes the scientists studying them are not sure whether they are looking at actual surface detail or featureless terrain. However, Earth-based telescopic observers looking towards Mars when its Martian atmosphere contained less dust and clouds could see the wave of darkening.

The dark area confirmed in 1952 was first observed in late 1948 and was situated northeast of the Syrtis Major. Earl Slipher of the Lowell Observatory observed a sharp darkening of this area in 1954 and it became referred to as "Slipher's spot" in Nodus Laocoontis (Laocoon Knot). Observations by Russian astronomers in 1956 showed the spot's shape as a triangle or trapezoid, with its sharp angle pointing toward the Syrtis Major region. The spot's color, estimated without a filter on September 7, 1956, was dark-green (The Astronomic Council Committee of Planetary Physics, Moscow, 1959). There were no major dust storms preceding this observation that could account for the development of this new dark area on Mars.

On January 16, 1950, a Japanese astronomer, Tsuneo Saheki, was observing Mars from an observatory near Osaka. Saheki described a strange cloud formation in Mare Chronium in the southern hemisphere of Mars. The cloud was high in the atmosphere of Mars and could be seen on the limb of the planet as it rotated on its axis. Saheki was a seasoned observer, having watched many cloud forms on Mars, but this one was different. It was yellowish-gray in color. Most of the clouds he had observed were either yellow dust or blue water-vapor clouds. He estimated the length of the cloud to be about 2,500 miles. Saheki had seen a similar cloud in 1933, but this one was enormous. Speculation was that it was a volcanic event.

Then on December 8, 1951, Saheki again reported that he saw a sudden brightening in an area known as Tithonius Lacus near the Martian equator. This is the area of the Tharsis volcanoes and giant canyon known as Nocus Labyrinthus. Saheki's observation lasted five minutes, with only a fading white cloud remaining. Other brilliant flares have been reported by astronomers viewing Mars, but the popular consensus is that they could be meteor events, since the *Mariner* and *Viking* spacecraft did not find evidence of volcanic activity. Or did they? There is evidence provided by *Viking Orbiter* images that volcanic activity on Mars may be still happening. I will tie this in with my discussion in Chapter Six of the Russian MARS-96 Mission.

WHILE VOLCANIC ACTIVITY ON MARS might indirectly suggest life could exist there, Gerard P. Kuiper was an astronomer who advanced its notion by examining terrestrial lichens. Kuiper, born in the Netherlands, came to the United States in 1933 to work at the Lick Observatory on Mount Hamilton in California. He was the first to detect carbon dioxide in Mars' atmosphere, the first to find an atmosphere around Saturn's largest moon Titan, and the founder of the Lunar and Planetary Institute in Arizona.

Kuiper was interested in the possibility of life on Mars, and with all the previous groundwork done by other astronomers in a search for chlorophyll without positive results, he decided to take a different approach. In 1954 Kuiper became interested in the study of lichens, the symbiotic algae and fungi that grow in harsh and mild climates on rocks, trees, and soil. Through a detailed spectrographic analysis, Kuiper discovered lichens do not show evidence of chlorophyll when looked at by infrared spectroscopy. This led him to examine different combinations of lichens and rocks on Earth, and he discovered that lichens along with oxidized volcanic basalt rock matched very closely the spectrum of the dark areas of Mars during the spring, summer, and fall seasons. Kuiper observed Mars during the favorable close approach (opposition) to Earth in 1954 with the 82-inch reflecting telescope of the McDonald Observatory in Fort Davis, Texas. The importance of this observation would later indirectly lead groups of researchers to study lichens in the Antarctic desert as possible Martian analogs of life. However, Kuiper stated:

> The hypothesis of plant life, for reasons developed elsewhere, appears still the most satisfactory explanation of various shades of dark markings and their complex seasonal changes... I should, however, correct the impression that I have supposed this hypothetical vegetation to be lichens. Actually I said: Particularly, the

comparison with lichens must be regarded to have only heuristic value; it would be most surprising if similar species had developed on Mars as on the Earth. *(Publications of the Astronomical Society of the Pacific,* Vol. 67, page 281, October 1955)

Into the Infrared

William M. Sinton became a Research Associate at Johns Hopkins University in 1953 after deciding to change his career goals from physics to astronomy. The reason Sinton made the switch was directly because of the influence of Gerard Kuiper and his work at the University of Texas regarding infrared studies of Mars. Sinton also admired the work done by pioneers such as Percival Lowell, Otto Lampland, and Earl C. Slipher. He saw tremendous potential for the new (in his era) and advancing astronomical techniques of spectroscopic and photometric measurements.

Under the tutorial direction of Professor John Strong of Johns Hopkins University Astronomy Department, Sinton wrote his thesis on the "Infrared Studies of Planets" in 1954. After graduation he began working at Harvard College Observatory, where he specialized in infrared technology, physics of planets, and astronomical instrumentation. Sinton's fascination with Gerard Kuiper's suggestion that Mars might have organisms similar to terrestrial lichens set him on a course of discovery that would remain controversial to this day—the search for life on Mars using the remote sensing technique provided by infrared spectroscopy.

Sinton said in 1996 of his quest:

> I guess I was really looking for a project that would make a name for myself! It looked like something that could be done. Along with the work of Lowell, Lampland, Slipher and Kuiper, I also read the books *Astrobiology* by G. A. Tikhov, *The Red and Green Planet* by Hubertus Strunghold, and of course *The Planet Mars* by G. de Vocouleurs. It made me think that if a planet had enough vegetation, even lichens on it to make it look green, then it certainly would have enough life to show organic absorption bands in the infrared. (personal communication)

DR. WILLIAM M. SINTON

Dr. William M. Sinton relaxes in his Arizona home in 1992. Sinton was the first to find spectroscopic evidence for "vegetation" on Mars in 1956.

Sinton knew that all organic molecules possess strong absorption bands near the 3.4-micron region of the infrared called carbon-hydrogen (or C-H bonds) resonance. He noted that light reflecting from plants penetrated beneath the surface before emerging and subsequently shows the absorption bands of the plant material. In organisms such as certain lichens or mosses which are adapted to dry environments, the absorption features are pronounced. Sinton knew that Mars had very little water, and that if the planet did indeed have something like plant life changing colors from green to brown seasonally, then it might all be similar to lichens and mosses. He proceeded to make spectrum analyses of terrestrial lichens and mosses to use as a comparison to what he might observe on Mars.

To begin his search for life on Mars, Sinton used a liquid nitrogen-cooled lead sulfide detector on a Perkin-Elmer Model 83 monochromator with a lithium fluoride prism attached to a 61-inch reflecting telescope in the Harvard Observatory. His observations started with the 1956 opposition of Mars in September and continued for a period of three weeks. These observations covered the entire disk of the planet at five different wavelengths per night in the 3–4 micron range. Sinton was astounded to find that Mars had a C-H bond feature at the 3.45-micron wavelength in the infrared portion of the spectrum which looked strikingly similar to the spectrum of Earth lichens! He published his results in 1957 and sent a shock wave rippling through the astronomical community.

Not satisfied with just one set of results that represented the whole disk of Mars, William M. Sinton thought he should use a larger telescope and look to see if there were differences between the light orange desert areas of Mars and the greenish dark areas of Mars. To do this Sinton obtained the use of the largest telescope in the world at the time, the Mount Palomar 200-inch reflecting telescope. This would provide ten times the sensitivity he achieved in 1956 at the Harvard Observatory.

Over the course of a two-week observation period during Martian opposition in 1958, Sinton made thirty-six disk-resolved spectra of the light desert and dark regions on Mars. Three distinct absorption bands were observed again at 3.45, 3.58, and 3.69 microns. Sinton thought that organic molecules were produced in localized regions in relatively short spans of time, and vegetation seemed a likely explanation for the appearance of organic molecules. Proof of this rapid change on the surface of Mars was impressed on him in 1954 by the large 580,000-square-mile area called Slipher's Spot. Sinton also knew that chloropyll was not the only pigment plants could use that absorb in the infrared, so previous searches for evidence of chlorophyll did not discourage him. The absorption band at the 3.69-micron range seemed puzzling because he could not find a terrestrial plant match-up. However, just as he was preparing to publish his results, it was discovered that a species of alga, *Cladophora*, had the ability to reproduce his findings on Mars.

Sinton now knew that the *Cladophora* comparison spectra revealed organic molecules, but what came as a surprise was the fact that carbohydrate molecules present in the plant could also be observed in the infrared. This indicated to Sinton that plant life on Mars might store larger quantities of carbohydrates (food) than terrestrial plants do. The Lowell Observatory astronomer felt it was also possible that

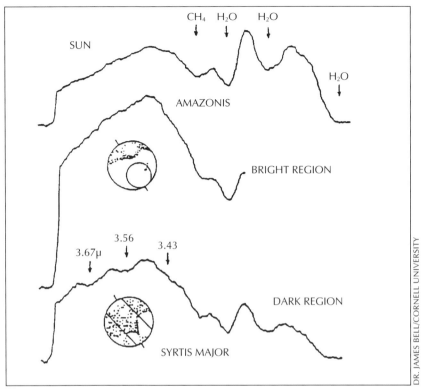

Graph depicting the Sinton organic bands for both the light desert regions of Mars and the dark maria.

his Mars spectra bore a close resemblance to carbonates (calcium carbonate), but concluded that organic molecules represented his findings more accurately. Sinton published his results in the November 1959 issue of *Science* magazine, with the title "Further Evidence of Vegetation on Mars."

Such controversial evidence always leads to a point and counterpoint interplay among research scientists. As soon as one researcher publishes a new finding, it is almost the duty of another scientist to try to disprove that new theory or show that it is something different but not unique. This laborious process is necessary to verify new discoveries; however, some scientists do use the genuine real discoveries of peers to make a name for themselves only. This age-old

79

push-and-pull concept of gaining notoriety in science is unfortunately very prevalent in today's fast-paced information age, and it does tend to slow down progress in research. A case in point is the varied studies on coffee and caffeine that keep coming out in medical journals—one study one week shows it's safe to drink, the next shows that it is not!

Sinton had his critics, of course, but one sharp researcher, N. B. Colthup of Stamford Research Laboratories in Connecticut, wrote a letter to *Science* in response to the "Sinton bands," as they were now being called. In his letter in the August 1961 issue of *Science*, page 529, Colthup makes the observation that organic aldehydes show in the infrared at around the 3.56, 3.53, 3.65, 3.67, and 3.70-micron bandwidths. He even postulated that acetaldehydes may be formed on Mars as a result of anaerobic metabolic fermentation of carbohydrates. Sinton responded to Colthup's letter favorably on the same page, stating his work was in agreement with Colthup's findings.

Even though it was known during Sinton's infrared studies of Mars that the planet had areas that appeared to be covered in a low-lying fog, it would not be until the mid-1980s that Russian researchers from the Bakh Institute of Biochemistry in Moscow documented the importance of fog in the formation of organic aldehydes. T. E. Pavlovskaya and T. A. Telegina noted that the ultraviolet light of the sun passing through fog with aldehydes, rain droplets, and clouds could produce amino acids including glycine—important precursors of life. If Sinton and Colthup were correct about aldehydes playing an important role on Mars, then combined with the ice-fog cycle on Mars, there would be yet another mechanism for the production of organic material in vast quantities over geologic time. The relevance of organic material on Mars would be the deciding factor for the Viking biology experiments in 1976. If there was organic material detected by the Viking GCMS organic analysis instrument, then Mars would be declared alive; if not, as it was claimed, then a dead planet.

The "Sinton band" controversy has never been fully resolved quantitatively, although the interpretation of two of these bands (3.58 μm and 3.69 μm) in 1965 by D. G. Rea, T. Belsky, and M. Calvin as being produced by terrestrial heavy water molecules (HDO) is generally accepted, though not completely proven. In the January 1965 issue of *Science*, pages 48-49, three University of California scientists from the Department of Chemistry published a letter entitled: "Sinton Bands: Evidence of Deuterated Water On Mars." The authors conclude that although Rea, Belsky, and Calvin's evidence best fits the HDO spectrum as opposed to organic acetaldehyde production, a question remained open. Since William Sinton had only observed his spectra in the dark regions of Mars, it must mean that water (HDO) is concentrated within these areas and should be a consideration for where to send a robotic spacecraft to search for life.

Rea, Belsky, and Calvin offered no explanation for the Sinton 3.45-micron band, and this is still very much an area of active research today. Although this wavelength in the infrared spectrum points to aldehyde and therefore organic material production on Mars (biology?), most scientists studying the issue today feel that because the Viking GCMS organic analysis instrument did not find evidence for organics on Mars, it is highly unlikely that this band can be attributed to organics.

But what if the Viking GCMS was unable to detect small amounts of organic material on Mars even though life might be present? There is strong evidence that this was the case, as we will explore. Sinton would later retract his findings on vegetation on Mars in a letter he sent to *Science* and agreed with Rea, Belsky, and Calvin that more than likely HDO was observed and not organics. However, it was later discovered that the Rea and O'Leary infrared spectra were actually spectra of Earth's own atmospheric HDO! Did William Sinton find HDO in Earth's atmosphere? If so, why did it only appear when he was observing the dark regions of Mars? Perhaps pressure from the scientific community was too much for Sinton in light of his controversial findings, and therefore he succumbed to "peer pressure."

I wrote to Dr. William Sinton in January of 1995, asking him if he was still convinced of the Rea, Belsky, and Calvin findings. He replied, "I hope that you will agree with my conclusion that there is no firm evidence for my earlier suggestion of vegetation on Mars."

Dr. James Bell of the Cornell University Department of Astronomy wrote to me in 1996 regarding Dr. Sinton and his spectra:

I think that William Sinton was impeccable in his technique and he was at the forefront of the field of IR spectroscopy when he took those Mars data in 1958 and 1961. While I think that his data are of high precision, I agree with others that the interpretation of at least two of the features he detected is related to Earth atmospheric absorptions rather than anything Martian. The third band at 3.45 microns remains enigmatic, however. My personal feeling is that this band may be related to a carbonate mineral or some other evaporite salt that preferentially occurs in the Martian dark regions: I base this on preliminary analysis that I have done on some new Mars 3–4 micron imaging spectroscopic data that I obtained in 1995 from Mauna Kea Observatory. These data show evidence for absorption features at wavelengths near the Sinton Bands and which occur primarily in the dark region Syrtis Major. These very preliminary results are awaiting confirmation by more detailed analysis underway. I think the likelihood of the 3.45-micron feature being due to an aldehyde is remote, especially since the Viking Lander GCMS data showed there to be little or no organic matter in the soils, and there is a very high UV flux from the Sun constantly bombarding the Martian surface. But the correspondence between the features we detected and the C-H bonds is tantalizing, though it points out we must be very careful to remove possible terrestrial atmospheric contamination. My collaborators and I are continuing to explore every possibility, as Mars is a very surprising place. . . .

Dr. Theodore L. Roush of the Department of Geosciences at San Francisco State University and his colleague Dr. Yvonne Pendleton of the NASA Ames Research Center Space Sciences Division in California are today's cutting-edge experts in the field of infrared spectroscopy. Their work has taken Dr. William Sinton's important concepts to the limits of the cosmos in the search for organic material. The Sinton Bands 3.4-micron feature has recently been detected in external galaxies, indicating the widespread availability of organic material for incorporation into planetary systems. Dr. Roush says of Mars and the Sinton Bands:

> A number of spectroscopic observations of Mars exist at pertinent wavelengths. Some of the more recent data should be

evaluated more carefully, focusing on the signatures of organics. Modern observational techniques can provide high signal-to-noise ratio data with better atmospheric correction... The Sinton Bands controversy was never fully resolved, although the interpretation of Rea in 1965 for giving rise to two of the bands is generally accepted. There are still remaining questions regarding the source of the third band, 3.45 microns (personal communication).

Dr. Sinton's pioneering contributions in the study of C-H bonds in the infrared have given way to a whole new branch of research looking for evidence of organic molecules in space. Some of today's most talented astrophysicists, including Roush and Pendleton (of the NASA Ames Research Center) and James F. Bell of Cornell University's Center for Radiophysics and Space Research, have been finding organic molecules in comets, meteorites, and interplanetary and interstellar dust. Evidence for polycyclic aromatic hydrocarbons (PAHs) has been discovered near the center of our galaxy. These PAHs have recently been found in meteorites—in particular, the Martian meteorite announced in August 1996 by NASA to contain microfossil evidence of life on Mars (ALH 84001). The production of organic material in the very deepest regions of space may indicate that life gets started in the vast clouds of interstellar gas and dust.

In 1982 astrophysicists Fred Hoyle and Chandra Wickramasinghe suggested that a C-H infrared stretching band of diffuse dust in interstellar space may be due to organic materials of biological origin. They demonstrated that the spectrum of *E. coli* bacteria closely matched the C-H stretching feature in interstellar and cometary spectra (Hoyle, F., Wickramasinghe, C., 1984). More recent discoveries regarding the organic infrared bands are that C-H stretching features are very similar to those produced when UV irradiation and the subsequent warming of interstellar ice analogs produce organic residues (Sanford, S.A., Pendleton, Y., 1991). Furthermore, UV irradiation of cometary ices was found when warmed to 150 kelvin to form more and more complex organic material as the original compounds are slowly evaporated into space (Allamandola, L.J., 1988). Organic material is found in the vacuum of deep space, in comets and meteors, in interstellar ices, and distant galaxies. Thus it is difficult to conceive of a Mars without organic material.

WHILE WILLIAM M. SINTON WAS carrying out his significant work, Jean-Henri Focas of the National Observatory of Athens was busy with work of his own during the oppositions of Mars in 1954 and 1956, but mainly those of 1958 and 1960–61. Focas concentrated his efforts on trying to resolve the "wave of darkening" phenomenon and began a systematic study of the dark areas of Mars from the Pic du Midi Observatory in France. This observatory has an altitude of 10,000 feet above sea level and offers some of the best viewing conditions in the world. Focas' work would incorporate more than eight years of telescopic and photographic observations of Mars with the observatory's 25-inch refractor, using powers on the telescope as high as 1200x! Truly extraordinary atmospheric clarity is necessary to view with powers of 1200 on any telescope. At times Focas had a resolution of objects on Mars as small as forty miles in size. In his study Focas found two separate waves of darkening alternately originating at both poles at half-martian-year intervals and spreading towards the equator (Focas, 1962). His conclusions were that the dark areas follow a seasonal cycle of water vapor release from the polar ice caps and that the intensity of the dark areas increased with waves generating towards the equator at an approximate speed of thirty-five kilometers per day.

This careful ground-based telescopic study of the seasonal wave of darkening by Focas would have its critics, but Focas was careful not to mention colors or possible Martian organisms in his study "Seasonal Evolution of the Fine Structure of the Dark Areas of Mars" (*Planetary and Space Science*, 1962, Vol. 9, page 371). By 1962 the colors of Mars were a controversial issue, as many astronomers who no longer believed the canals to be actual features of the planet also attributed the green areas of Mars to optical illusion caused by contrast of the orange desert regions with what was thought to be gray dark regions! Many astronomers still dispute the colors of Mars, and with the confusion of the Viking color calibrations on Mars, it's no wonder. For those doubting the green color of Mars as a real phenomenon, just look at the photograph released by the Hubble Space Science Institute: Photo Release Number STScI-PRC95-17B, entitled "Springtime On Mars: Hubble's Best View of the Red Planet," taken February 25, 1995. Taken at a distance of more than sixty-five million miles away, the Hubble Space Telescope photograph shows white and blue sky clouds with hazes, a brilliant white polar ice cap, orange deserts, and dark green areas! So much for the illusion, but if you do not feel Mars can sustain life—as many scientists do not— then green cannot exist either. So rather than admit they see this color, they choose to believe it to be a contrast effect. Remember my story of Ferdinand Magellan at Tierra del Fuego?

The great discoveries of the early microbiologists, chemists, and astronomers led to a merging of sciences we call exobiology today. With the arrival of the spacecraft era and the launch of Russia's first satellite, *Sputnik* in 1957, astronomers, geologists, chemists, and microbiologists knew what would come shortly—the robotic exploration of the planets by spacecraft. Of course, because of all the work done previously by researchers on Mars and possibilities of life existing there, it would become the target of a host of spacecraft from the Untied States and Russia. However, only three men would actually have one of their own experiments search the soil of Mars for a chance at finding life. That chance came with the Viking Mission to Mars.

The names of the chosen few were Dr. Norman Horowitz of the California Institute of Technology, Vance Oyama of the NASA Ames Research Center in California, and Dr. Gilbert V. Levin of Biospherics Incorporated in Maryland. Together these researchers would challenge the most cutting-edge knowledge of life on Mars, and one of them through the course of his perseverance would announce that "we" had "more likely than not" discovered living organisms on the surface of Mars in 1976, and he has maintained that position until today in spite of peers not wanting to examine his data closely, and in spite of their unscientific ridicule.

The next chapter contains a number of elements necessary to demonstrate how our society conducts planetary research, sets requirements regarding quarantine and sterilization procedures for planets and spacecraft, and how research in the harsh land of Antarctica played a key role in the discovery of life on Mars. It is also the story of how Dr. Gilbert V. Levin tested his sensitive instruments on Earth—showing their incredible range for the detection of microorganisms—and applied them successfully to the search for life on another planet.

CHAPTER THREE

A Red Planet on the Great White Desert

Many Antarctic soils contain small numbers of bacteria per gram, but it is doubtful if they are living there. There are rich sources of bacterial life in and around the Antarctic valleys—the ocean for instance—and there is every reason to believe that organisms get blown around by the wind and scattered over the valleys. In this respect, Antarctica and Mars are very different places.

—Dr. Norman H. Horowitz (1996), Emeritus Professor, CIT, and Principal Investigator of the Viking Pyrolytic Release Experiment

I N 1957, *Sputnik* WAS LAUNCHED into orbit and the exploration of space officially began. Several significant events would occur as an indirect result of the new spacecraft era. In this same year (1957), the U.S. National Academy of Sciences expressed concern over the possibilities that planetary exploration by spacecraft could contaminate the planets; by 1958 it assigned a special task force to the problem. The Committee on Contamination by Extraterrestrial Exploration (CETEX) was concerned with the contamination of the Moon, Mars, and Venus by Earth microorganisms. Already on the NASA drawing board were space probes designed to crash into as well as orbit the Moon: the *Ranger* series and *Lunar Orbiters.* Also planned were flybys of Venus and Mars with the *Mariner* probes. Russia of course had its own plans to explore the planets, and would in the end ignore the planetary sterilization requirements set up by an international committee.

CETEX also began to look at the Panspermia hypothesis as a possibility for contaminating Earth through a space vehicle picking up contaminated space dust on its exterior and returning the "germs" upon reentering Earth's atmosphere. The CETEX meetings led directly to the establishment of a worldwide space research organization in October of 1958—COSPAR, the Committee on Space Research, whose charter included studying methods for sterilizing the exterior and interior of international spacecraft in a way that would not harm the delicate scientific instruments inside. At the time of the Ranger class of moon probes, the methods of sterilizing spacecraft were thought of in hospital terms such as autoclaving with wet steam at 25 psi in small pressure chambers.

In 1959 NASA set up an internal biological group, and this ad hoc Bioscience Advisory Committee—known alternately as the Kety Committee (after its Chairman, Dr. Seymour S. Kety)—had as an advisor Dr. Joshua Lederberg (who coined the word "exobiology" in 1960). Lederberg recommended to NASA the U.S. Army BioLab division, which had extensive experience in the sterilization of large vehicles such as Army trucks and delicate laboratory equipment. This association between the U.S. Army BioLab division and NASA would prove useful in determining what techniques could be employed. The Army BioLab had experience dating back to World War II using various sterilization techniques, and they had perfected the use of a volatile noncorrosive gas: ethylene oxide. However, the gaseous treatment could only be used for the exterior of spacecraft. Joshua Lederberg knew enough about microorganisms to realize that even spacecraft screws might protect them—between the threads! He also realized that plastics and other materials housing wiring and other spacecraft components could shield microorganisms from the gas sterilization treatment. The Army BioLab study yielded the determination that dry heat sterilization at 125 degrees Celsius for 24 hours would be required to "sterilize to the extent feasible." In other words, the complete elimination of Earth-contaminating organisms could not be accomplished.

COSPAR also directly implemented studies of microorganisms under different planetary conditions to find out whether or not spores of bacteria could survive in the vacuum of space; to check the resistance of bacteria and other microbes to intense ultraviolet light; and most urgently, to determine whether or not Earth microorganisms could survive on Mars. COSPAR's other function was to serve the need for international regulation and discussion of other aspects of satellite and space probe programs. It reported directly to the International Council of Scientific Unions, to which all countries in the world that engage in scientific research belong.

THE SECOND SIGNIFICANT EVENT OCCURRING during this time period happened in Washington, D.C., at the signing of the Antarctic Treaty Conference in 1959. Twelve nations signed the treaty, including the United States and Russia (USSR then), with the express purpose of preserving the Antarctic environment as a place for scientific study. Although military logistical support is permitted by this treaty, military operations in Antarctica are forbidden.

Antarctica, because of its unique and extreme environment, holds great interest for biologists studying flora and microorganisms that can grow in a dry, windy, and cold region resembling a 450-million-year-old Earth, before large animal and plant life proliferated. Of course, to scientists wishing to understand what kind of life could survive on Mars, Antarctica would become the best available natural laboratory, serving as a microbiological testing ground starting in the austral summer of 1966, when a group of American scientists began to gather soil samples for NASA's pre-flight testing of the biology experiments for the Mars-Viking program.

In the dry and ice-free valley of Antarctica's South Victoria Land, evidence of microbes surviving under the harsh conditions of a Mars-like environment would be found, but surprisingly not by the scientists sent there on the NASA contract. It would be the result of the dedicated work of an enthusiastic microbiologist named Wolf Vishniac from the University of Rochester in New York, along with the support of Dr. Gilbert V. Levin and his co-experimenter in the Viking program, Dr. Patricia Ann Straat. This is the story of how this strange land covering over six million square miles played a crucial role in the discovery of life on Mars.

GILBERT LEVIN WAS A BUSY man in 1958. He was Vice President of a Washington, D.C., research and development company called Resources Research Incorporated, as well as Clinical Assistant Professor of Preventative Medicine at the Georgetown University School of Medicine and Dentistry. He also worked in the D.C. Department of Public Health. The tall, thin, and youthful environmental engineer had earned a Bachelor's degree in civil engineering and a Master's degree in sanitary engineering from Johns Hopkins University in 1948. Levin went on to obtain his Ph.D. in Environmental Engineering in 1963, also from Johns Hopkins University.

It's not surprising that a career in sanitary microbiology would lead Levin to the surface of Mars in 1976. In fact, it seemed like a natural extension of the work he had been doing in his field of public health for the states of Maryland and California, where he held the position of Public Health Engineer. Levin's expertise in the fields of bacteriology, mycology, phytology, virology, protozoology, and zoology made him an attractive NASA candidate for life detection on another planet.

Levin routinely worked on problems involving the determination of the bacteriological quality of water. While he was in the D.C. Department of Public Health, Levin invented a technique using miniscule amounts of radioactive carbon-14 to rapidly detect microorganisms in water. He convinced the Director of Public Health and the Dean of the Georgetown School of Medicine and Dentistry to let him work one day a week at the university to develop his invention. This technique provided radioactive compounds to media suspected of containing microorganisms and then looked for radioactive carbon dioxide gas to come off as evidence of their presence. It was also an excellent means of studying a microbe's metabolism. The microbial radiorespirometry technique explored by Gil Levin became the basis of his extraterrestrial life-detection experiment called the Labeled Release, which in 1976 would find evidence for life on Mars. Levin's unique idea came to him in 1951 and evolved through several changes before leaving for Mars. The early field version of this experiment would eventually be taken to Antarctica, to be used by Dr. Wolf Vishniac.

Gil Levin was ten years old when his first cousin, Evelyn Glickman, who was visiting him at the time, asked if he would like to go out in the backyard to look at the stars. Levin's cousin was attending Goucher College (now Goucher University) in Baltimore, Maryland, and had been taking an astronomy course while working on becoming a school teacher. Sitting on a bench in the backyard, Gil Levin got his first astronomy lesson. After identifying some constellations and stars, Levin's cousin pointed to a conspicuous orange-red object in the sky. She told him it was the planet Mars, and explained that some astronomers thought there could be living things on Mars. Mars would have to sit in the back of Levin's mind until he was married and had two of his three children. While these two children, Ron and Henry, were still very young, Gil Levin bought them a special present—a large backyard telescope which Levin used to teach them astronomy basics. Whenever Mars was available for viewing, it seemed

91

the Levin family could not resist its charm and allure. About three years after her birth, daughter Carol joined in. They viewed and witnessed for themselves the bright polar ice caps of Mars and dark surface markings.

By this time Gil Levin was well into his career in environmental engineering. While the stimulus of Levin's exposure to Mars through his cousin and later the family telescope would gain him an appreciation for astronomy and Mars, it wasn't until he tried to market his invention for detecting microbes that he would eventually go to NASA with it. Levin recalls:

> What got me to go look for life on Mars was actually my frustration in trying to promote use of this method on Earth. The technique was exquisitely sensitive in finding microorganisms, but I could not interest authorities in the health field because it contained a miniscule amount of a radioisotope."

Levin submitted one proposal for its use that was reviewed by a distinguished microbiologist at the Jewish Hospital in Brooklyn. The microbiologist wrote a letter to the agency Levin had applied to stating that the radiorespirometry technique should not be considered because of the radioisotope in it. The letter concluded with "we are moving away from the use of radioisotopes." Since that time, of course, hospitals are using far greater quantities of radioisotopes than ever. Ironically, when Levin's patent expired on the device, its technology became widely used in hospitals all over the world for the quick detection of blood infections.

Nevertheless, Gil Levin had perhaps the most sensitive microbial detection device in the world, and he thought, "If they felt it was not safe enough to use on this planet, perhaps NASA might consider it for another world." As a biochemist, Levin was interested in the origin of life and became well versed in the Oparin-Haldane theory. In fact, he met A. I. Oparin once and had him sign a copy of his famous book *The Origin Of Life*. Levin, though, had his own ideas about how life may have developed. He became so intrigued with the problem he began to write a paper on the subject in the early 1960s, which was never published because of discouraging remarks made by Dr. Norman Horowitz of CalTech, who at the time was assigned by NASA to work with Levin because Levin still did not have his "ticket" (Ph.D.). Horowitz told him, "You should never publish this, it's ridiculous." Levin respected Horowitz's opinion at that time and filed the paper away until I requested to use it for the book now in your hands. Presented here for the first time are excerpts from Levin's original unpublished paper that clearly demonstrate a model of pre-biotic chemistry that is current with many of today's latest theories, but was written more than thirty years ago:

> An apparent anomaly exists between currently developing theories to explain the origin of life through chemical evolution... The anomaly in the theory of chemical evolution is why more than one biochemical form of life did not develop on the Earth? We prepared to explore Mars for life on the basis that it possesses an environment roughly similar to that on Earth. We speculate on whether, if life is found on Mars, it will consist of basic building blocks other than the 20 amino acids used by Earth life ... and whether its genetic message will be transmitted by other than DNA. Yet, on Earth, where we know the environment is favorable for life, all known life is extrapolated to a single line of descent. If life is the inevitable result of chemical evolution, why did it not arise in more than one biochemical form on Earth? Darwin recognized this problem and rendered the explanation that, if a second biochemical species were to appear, it would instantly be devoured by the life forms already in existence. This explanation, however, presumes that the initial life forms were distributed to all possible niches where the second form might arise prior to the latter event. The explanation also assumes a greater degree of predation of one biochemical species upon another (perhaps not even nutritive to the first) than we observe between the highly competitive but, nonetheless, continuously surviving strains of the same species....
>
> ... There are other speculative ways why apparently only one genesis has occurred on Earth despite the inexorable transition of chemical evolution into biological evolution in accordance with the increasing acceptance of the latter theory as indicated above. One method of severely limiting the inevitability of the statistical

route to genesis would be the requirement of forces or conditions available only in the naked space environment at some intermediary point in pre-biological evolution. For example, some of the initial life precursor reactions might require the planetary environment such as a reducing atmosphere, gases, and water. Perhaps subsequent steps in the evolutionary chain might require ultraviolet or ionizing radiation intensities not attenuated by the planetary atmosphere, low temperature, and desiccation. A return to the planetary environment might be necessary for completion of the remaining steps producing living material. Thus, for example, chemical evolution may begin on Planet A and progress to some point beyond which conditions are not favorable. Meteoritic impact or other catastrophic force might cause planetary material, including the various significant evolutionary compounds, to be injected into space. The space environment might then promote additional reaction or effect necessary preparations for such reactions. Upon capture of this material by Planet B, possessing the necessary environmental characteristics, the evolutionary reactions might proceed toward the creation of life.

The reasons for suggesting this admittedly speculative hypothesis are not confined to the statistical effect, which would be to limit the evolution of life toward conformance with the observed frequency while still admitting chemical evolution. Several factors in the space environment, singly or in combination, may be required for or may assist in various reaction steps in the life precursor sequence. The principal source of energy for abiogenic formation of biological compounds generally cited is the ultraviolet portion of the spectrum. Proto-planetary atmospheres and planetary atmospheres might be expected to attenuate severely the UV flux. Thus, much higher energies are available in space. This, of course, is also dependent on the location of the reacting substances in space with reference to a stellar source of UV. Similarly, higher incident energies from ionizing radiations are available in space than on planetary surfaces. Such radiations are also postulated as energy sources for chemical evolution. However, the low temperature and lack of water operating on the material in space may be the most important factors. Free radicals would be produced by UV irradiation of the life precursor compounds.

Accordingly, the free radical mechanisms may play an important role in chemical evolution. However, since free radicals are transitory in nature, under planetary conditions free radicals would have very short lifetimes and the population of available species would always be low. Accordingly, opportunities for particular combinations might be severely restricted. However, recent technology has devised the means for the preservation and storage of free radicals for subsequent reaction. The radicals are stored in a frozen state, such as on a cold finger. The low temperature and desiccating conditions in space might similarly serve to accumulate free radicals produced by UV flux. Upon re-entry into a planetary atmosphere, the material would rapidly warm and an explosive-like reaction of the now-liberated free radicals would occur. The high populations of a great many species of free radicals which would have developed during the space sojourn might be expected to produce a very fertile reaction mixture, out of which might come some compounds of great, if not singular, importance in the evolutionary theory.
—Gilbert V. Levin, November 11, 1964

Levin's concept—that the origin of life might have its beginnings in the cold of space—was not easily accepted in 1964. Extraterrestrial origins of life and Panspermia were refuted soundly by most biochemists studying the origin of life, particularly those who had theological attachments to the "creation" of life as the exclusive providence of Earth. As surprising as this may seem, some prominent scientists today maintain that Earth is the only planet that can have life, based on "holy scripture."

ONE NIGHT AT A WASHINGTON, D.C., cocktail party in 1959, Gil Levin met NASA's first administrator, Dr. Keith Glennan. Levin asked Glennan if NASA had any plans to go to Mars and look for life. The two discussed Mars as the logical first target for extraterrestrial life-detection experiments. To Levin's surprise, Glennan told him that NASA had just hired someone to head its new biology branch—a Professor of Neurology from Western Reserve University named Dr. Clark T. Randt, who went on to become first director of NASA's Office of Life Sciences. Glennan suggested Levin see Randt.

Randt listened to Levin's ideas on detecting microorganisms using his radiorespirometry technique and suggested that he submit an official proposal to NASA. Randt also informed Levin that a life-detection experiment had already been funded. When Levin asked whose, Randt told him that Dr. Wolf Vladimir Vishniac from the department of microbiology at Yale University was the first to get funded for a method for the detection of microorganisms for a future Mars probe. The device he invented was called the Wolf Trap and later called the Light Scattering Experiment.

Levin immediately submitted his proposal and was asked to come to NASA Headquarters in Washington, D.C., to discuss the details for funding of his experiment. While he was there, Levin met Wolf Vishniac and the two scientists talked at length on the subjects of Mars, the origin of life, and exobiology. Vishniac told Levin he had been studying algae, molds, and bacteria for ten years while at Yale. Vishniac's main thrust while studying microorganisms had been the chemoautotrophic bacteria such as Thiobacillus and other iron- and sulfur-reducing bacteria which he thought would be good candidates to look for on another planet. Of course, Levin explained his radiorespirometry technique and the sensitivity it had for the detection of small colonies of microorganisms. Vishniac then explained the concept of his experiment, the Wolf Trap (Light Scattering Experiment):

> In my experiment I can see if any microorganisms are reproducing by administering a small (1.5 cc) Mars soil sample to a stainless steel cup containing distilled water inside the experiment chamber, where it will be monitored by a sensitive optical detector. The inside of the test chamber is illuminated by a small lamp. If there is something growing in the solution, the water-soil solution will turn from clear to cloudy. This of course assumes that Martian microorganisms will grow in water. As a control, a clouded sample will be heated to 160 degrees Celsius for 15 days, killing any organisms in the solution.

The meeting between Levin and Vishniac led to a respected friendship and later a collaboration when Vishniac took his Mars experiment and Levin's to Antarctica.

It was both a very exciting and frustrating time for Gil Levin. NASA, COSPAR, and a host of other organizations and committees were meeting frequently. In 1961, Dr. John Olive, then Executive Director of the American Institute of Biological Sciences, appointed Levin to a committee for the purpose of examining the possibilities for biological experiments in the newly available environment of space. The title of that particular meeting was "Space Biology." About fifteen prominent scientists were in attendance, including Dr. Dean Cowie, a physicist of the Carnegie Institute in Washington, D.C., and Dr. Philip H. Abelson, editor of the prestigious journal *Science*. Levin recalls that meeting:

> Phil Abelson had come with Dean Cowie to our meeting, and when it was clear that what Dr. John Olive was talking about was the possibility of life on other planets, Phil Abelson stood up and grabbed Dean Cowie by the hand and said, "Let's get out of here, this is talking about looking for life on the other planets! The Bible tells us there cannot be any life on other planets—this is a waste of time!" And the two departed. I was shocked, we were shocked. Later, it was under Abelson as editor that *Science* rejected two of our scientific papers on the possibility of having detected life on Mars.

While still working as Vice President of Resources Research Incorporated in 1961, Gil Levin began building a life-detection experiment that could be bolted to a spacecraft, survive the launch, the interplanetary cruise, and entry into the Martian atmosphere, then survive the jolt of landing and function on the surface of Mars. This challenging task faced all those who dared to make a proposal and win a NASA grant. By 1962, the NASA Office of Life Sciences did not like the idea that Levin was not a Ph.D. yet. So as a condition of his accepted proposal, NASA asked him to pick a NASA-approved Ph.D. biologist to work with him on his life-detection experiment. Levin was given several names along with their credentials, and after careful consideration, he selected Dr. Norman H. Horowitz from the California Institute of Technology, who also was the Bioscience Section Manager at JPL and worked on a planetary quarantine committee with such notables as Dr. Carl Sagan and Dr. Joshua Lederberg. Levin says,

> NASA explained to me that since I did not have a Ph.D. at the time, it would be desirable to have someone with a union ticket work with me in case I had a successful experiment, so that it could be presented, for example, at the National Academy of Sciences by a Ph.D. Those were the words the NASA official used. Horowitz agreed and visited our lab for a day about once every three months.

In 1963, Levin went to work for Hazleton Laboratories Incorporated of Falls Church, Virginia. Here he served as the Director of Special Research and was also Director of its Life Systems Division, which he established. The division initiated new bioengineering research efforts and product development through the company's scientists and engineers assigned to it. There Levin continued work on his radiorespirometry extraterrestrial life-detection experiment which

he called Gulliver, after the fictitious hero of Jonathan Swift's novel *Gulliver's Travels*. Levin thought the name might prove appropriate if his instrument travelled great distances and discovered strange, Lilliputian life forms. Gulliver would go through several transitional phases, with three versions fully developed and tested. Dr. Norman Horowitz would review the work and make some suggestions for experiments every time his schedule required him to drop by Hazleton Labs. However, the invention was Levin's, and Levin knew no one could understand it better than he. Horowitz was amazed by its simplicity of design and even more by

Dr. Norman Horowitz, in 1976. Principal Investigator for the Viking Pyrolytic Release experiment.

DR. GILBERT V. LEVIN

DR. GILBERT V. LEVIN

"Gulliver" became the NASA Exobiology Program centerpiece in the 1960s. It was later revised and developed into the Labeled Release experiment that went to Mars in 1976.

its exquisite sensitivity. Levin had a winner here, and Horowitz knew it.

The concept of Gulliver was to launch two small projectiles 25 to 100 feet from a landed Mars vehicle. Each projectile was attached to a "string" or line coated with silicone grease. Each string would then be reeled back into its test chamber. As it was being dragged through the Martian soil, particles would stick to the greased string. Once inside the chamber, the soil would be dosed with a tiny drop of radioactive compound. If any microorganisms were present which, like Earth organisms, could use the compounds as food, they would produce radioactive breath. This "tagged" gas would rise up and be detected by a Geiger counter. The radioactivity was continuously measured and the data would be relayed to Earth for analysis.

While Levin perfected Gulliver, other events were unfolding that demanded attention. Coming up in the fall of 1964 were three space probes going to Mars—*Mariner 3* and *4* from the United States and *Zond 2* from Russia. Both Gil Levin and Wolf Vishniac were extremely concerned about the Russians complying with COSPAR planetary sterilization guidelines. Relations between the U.S. and

DR. GILBERT V. LEVIN

A 1956 photograph showing Gil Levin (center) with Dr. Walter Hess (right) and Venton R. Harrison, all of the Georgetown University School of Medicine where Levin began his carbon-14 radiorespirometry work.

Russia were at an all-time low during this period, with both countries still reverberating from the Bay of Pigs incident in Cuba. Perhaps also there was a certain element of Russian arrogance in being the first to put a space vehicle into orbit. Did the Russians think they had a first claim to the dominion of outer space? Levin announced at a variety of committee meetings that the Russians did not seem concerned enough with the forward contamination issue and feared they would contaminate Mars with one of their space probes. A report coming out of COSPAR that shook everyone up stated that terrestrial microorganisms could produce an irreversible scientific catastrophe on Mars. In other words, Earth organisms deposited by an unsterilized spacecraft could overrun the Martian environment. To put it into terms everyone could understand, the researchers claimed that if the Earth were a sterile world, it would only require a few months or years for the entire planet to be populated with the descendants of a single cell! To illustrate the problem even more dramatically, it was calculated that if you took two grams of *E. coli* bacteria

DR. GILBERT V. LEVIN

Dr. Levin and other scientists test Gulliver in Death Valley, California, in 1964.

with a generation interval of only 30 minutes, they would, under ideal conditions, require only 66 hours to equal the Earth's mass (Davies, R. W., and M. G. Comuntzis, 1960).

Thus, without saying it, the researchers postulated a situation that if reversed would pose a back-contamination emergency of unparalleled magnitude for our planet if a Mars sample were returned with organisms that found our world "easy living."

On November 5, 1964, *Mariner 3* was launched with Mars as its goal in a first attempt at a flyby—past the planet, not entering orbit. This was considered a relatively safe method of exploring the planets without contaminating them. However, *Mariner 3* failed shortly after launch. Again the U.S. tried for Mars on November 28 with *Mariner 4*. This very successful mission eventually returned twenty-two photographs at a minimum distance of 6,118 miles above the Martian surface in 1965, but found no evidence of life.

However, Levin and Vishniac's worst fears were realized shortly after the launch of *Mariner 4*. On November 30, 1964, Russia launched the *Zond 2* spacecraft to Mars. Before the space probe completed its mission all communications were lost with *Zond 2* around May 4 or 5, 1965. From all indications of the vehicle's trajectory, it would impact Mars unless communications could be re-established and a rocket firing implemented to push the vehicle into a flyby mode. To this day, no one knows for sure, but it is believed that *Zond 2* crashed onto the surface of Mars in 1965. Ironically it carried a metallic bust of Lenin to commemorate the first Mars landing.

The Boeing Corporation of Seattle, Washington, conducted a study in the 1960s on the release of viable organisms which demonstrated the survivability of microorganisms impacting a surface at high speed. They put viable spores of bacteria in plastic bullets and fired them at different speeds into a variety of surface materials, including a simulated Martian environment. They were astonished to find that many organisms survived the impact. Their recommendation to the COSPAR committee based on the study was that encapsulated microorganisms must be killed prior to launch in order to reduce the risk of contamination of Mars or other planets being studied by spacecraft.

The 1960s were an incredible period of testing and research into the nature of microorganisms and their adaptations to extraterrestrial environments. A number of companies were contracted by NASA and the National Science Foundation to find out how tough microorganisms are. From 1961 to 1969 the Illinois Institute of Technology Research Institute was directed to study terrestrial aerobic and anaerobic microorganisms in extraterrestrial environments. This company exposed Earth organisms to a simulated Martian environment that included low temperatures, low pressures, high ultraviolet light intensities, and so forth. Their findings yielded evidence that microorganisms could not only survive on Mars, but in space as well!

The Massachusetts Institute of Technology in Cambridge investigated from 1962 to 1968 how Earth microorganisms are affected by space's ultra-high vacuum temperature, ultraviolet light, and ionizing radiation. Also studied were the effects of dry heat resistance of microbial spores as influenced by external or internal moisture. It was found that bacterial spores in a freeze-dried state, or "lyophilized,"

could lie dormant indefinitely and be revived to metabolize immediately upon the introduction of liquid water. This discovery would
have profound implications for a planet like Mars, with its atmospheric water vapor content and possible thin films of water in its soil,
along with water-ice. The Space Science Board appointed by the
National Academy of Sciences in 1963 summarized its interest in
Mars by giving the "highest priority to the prevention of the biological contamination of Mars until sufficient information has been
obtained about possible life forms so that further scientific studies
will not be jeopardized." Other studies under simulated Martian conditions showed that some Earth soil bacteria not only survived, but
multiplied. Spore-forming bacteria were found to have the highest
rate of cell growth as opposed to vegetative cells such as algae, which
did not produce spores for survival under extreme conditions. Most
interesting of all was the discovery that some Earth soil bacteria preferred the varied temperature cycles of Mars as opposed to comfortable Earth-like room temperatures!

In all, about twenty-one major organizations including the U.S.
Food and Drug Administration participated nationally in the microbial survival studies, along with investigations on how to better and
more economically sterilize spacecraft.

In 1963, Gilbert Levin obtained his Ph.D. in environmental engineering from Johns Hopkins University, having gone to school full-
time for three years while he still worked three-quarters time at
Resources Research, Inc. With that event, his life and work were
about to make an interesting transition. Levin knew it was just a
matter of a year or two before NASA would solicit a biology team
for a Mars-bound spacecraft, and so he intensified his work on Gulliver and also began to develop and test a new life-detection experiment. He called it the "Dark Release" experiment.

When Norman Horowitz showed up at Hazleton Labs for his
usual once-every-three-month check-up on Levin, he was shocked
to find Levin busy on yet another experiment evolved from a suggestion Horowitz had made to Levin as the result of his work on

Gulliver. Soon after Gil Levin gave him a demonstration of the Dark Release experiment, Horowitz announced he was going to leave the collaboration. Levin wondered why Norman Horowitz picked this particular time to leave, and although Horowitz did not mention why, it became apparent to Gil Levin in the months ahead.

Levin's novel new life-detection instrument, the "Dark Release" experiment, was designed to look for Martian photosynthetic organisms, should they exist. Levin explains how it works:

> You take a soil sample which you suspect might have something like algae in it and put it in a small cup in a closed chamber. Then you introduce into the chamber radioactive carbon dioxide, turn on a small lamp simulating sunlight for fifteen minutes, then turn off the light and with pressurized gas blow out the excess radioactive carbon dioxide. Then you put a "getter" [explained in detail later] pad at the top of the test chamber and keep the chamber in the dark for another fifteen minutes while the getter gathers any radioactive carbon dioxide that was expired by the algae. While in the light, algae pick up carbon dioxide; while in the dark, they release it. Algae in the light make glucose out of carbon dioxide and store it. When the light is turned off, the algae have to use the stored glucose for metabolism. An end product of that metabolism is carbon dioxide, which now is "labeled" from the radioactive carbon dioxide the algae incorporated in making the glucose. My idea was that by turning the light on and off, you could modulate the carbon dioxide being taken up or given off by the algae and then measure the response as time goes on.

After developing the experiment on soil naturally containing algae, Levin set up a demonstration for NASA officials who had already given the go-ahead for its use on a future Mars mission. Levin states:

> They were supporting it thinking that it was likely to develop as a complement to Gulliver. Gulliver would seek heterotrophic life while the Dark Release would seek photosynthetic life. Unfortunately, when I invited NASA down for the demonstration we had a guy freshly out of college who put on the demonstration

for us, and it didn't work! NASA left before we could repeat the experiment for them. When we later ran the exact same demonstration again, but collected the gas correctly, it worked beautifully. I submitted a detailed report to NASA showing that and explaining what had happened, but it was too late.

Levin says he later found out that Dr. Norman Horowitz had been funded by NASA for an experiment he called the Pyrolytic Release experiment (also referred to as the Carbon Assimilation experiment). When Levin had a chance to look over its design concept, he was stunned to find that it was an altered version of his Dark Release experiment.

Horowitz did not rely on the dark respiration to release the carbon dioxide from the soil sample after the carbon dioxide had been "fixed" in the light; he heated the soil sample and swept it with pressurized gas to drive out any radioactive carbon dioxide that had not been fixed. Then he raised the temperature to burn up, or "pyrolyze," whatever was in the chamber, and measure the radioactive gas coming out. He said that was the carbon dioxide that had been fixed and was evidence for life. He felt that in my Dark Release experiment where you turn the lamp on and off, some carbon dioxide would have been absorbed onto the soil and might come out interfering with the test. He wanted to drive all that out by a low level of heat literally forcing it out, then rely only on the "burning" as proof of what was bound by the organism. Nonetheless, I think the Dark Release experiment had the advantage because when you turn the light on and see the carbon dioxide level go down, then turn the light off and the carbon dioxide comes back out, this can only be from metabolism, not from soil absorption.

Probably the most incredible fact surrounding Horowitz and the Pyrolytic Release experiment was that it hardly ever got a response from an Earth soil, while Levin's Dark Release experiment routinely monitored metabolism of photosynthetic organisms in a variety of soils. In the test program Horowitz produced a response by growing cultures of algae, straining and filtering them out, in order to put a mass of algae in his experiment. He turned his lamp on inside the chamber and waited for the algae to pick up the carbon dioxide. He then turned the light off and finished the experiment. In this way, Horowitz could get a positive result. When Levin confronted Horowitz about the PR frequently not being able to get a response from Earth soil, he said, "The PR experiment was designed to look under Martian conditions and you should not be disturbed because it did not get a response from Earth soil."

On To: Antarctica

One of the most important discoveries made by the few scientists studying life in Antarctica in 1963–64 was the abundance of lichens growing on the rocks. Lichens are a community of algae and fungi living symbiotically, sometimes with bacteria, although less commonly. The algae harvest the Sun's energy through photosynthesis, and the fungi or bacteria live off the organic matter made by the algae and provide a protective structure for the algae. Most lichens live on land; however, there are some that live in water. Together they help each other survive extremities of moisture, temperature, and various levels of ultraviolet light intensity from the Sun.

When you recall from Chapter Two how the lichens were thought by astronomers Gerard Kuiper, William Sinton, and others to be good terrestrial candidates for life on Mars, then the study of lichens in the Mars-like environment of Antarctica seems essential. It was found that of the four hundred different lichen species in Antarctica, most preferred to grow on rocks larger than one inch in diameter with slope angles of less than 37 degrees. Lichens were not found on steeper slope angles. Lichens in an arid climate where the average daily humidity seldom gets above 10 percent were a fascinating find. How were these organisms getting their water? Was it possible that they lived on just the small amount of water vapor in the atmosphere? Other investigations into the nature of lichens found that they could actually live under one to fifteen inches of snow, and that this environment would moderate levels of light intensity and melt-water for metabolism. While the air temperature in Antarctica rarely gets above freezing, the temperature of the rocks (just as on Mars) often exceeds 100 degrees Fahrenheit! However, these high rock temperatures were found to dry out the lichens rather than help growth. The reason the lichens grew on north-facing rocks with slope angles of less than 37 degrees was to avoid direct sunlight and excess heating.

These early studies increased interest among scientists wanting to study microenvironments, and in particular those who wanted to study ecological and microbial systems that might be similar on Mars.

Dr. Roy E. Cameron (a microbiologist from the Darwin Institute), Dr. Jerry S. Hubbard, and Dr. Norman H. Horowitz, all working with the NASA Jet Propulsion Laboratory in Pasadena, California, carefully laid out plans to study the ice-free dry valleys of Southern Victoria Land, Antarctica. It was Horowitz who first suggested and encouraged studies in the Antarctic dry valleys because of their value to exobiology and Mars.

Cameron himself, with a number of colleagues under a grant from the Office of Polar Programs within the National Science Foundation, left for Antarctica in the summer season of 1965–66 for a three-year expedition. The trip to the "Great White Desert" continent took eight days. The scientists left California on a U.S. Navy LC-130F Hercules aircraft that had stop-overs in Hawaii, Guam, Australia, and New Zealand before finally reaching McMurdo Station, the home base for the United States Antarctic Research Program.

McMurdo Station had everything a scientist could need, including a state-of-the-art biological laboratory. Energy and heat were provided by a combination of a small nuclear power plant and diesel fuel. One of the advantages for scientific study in the Antarctic is you have sometimes four to six weeks of continuous daylight. This not only increased the amount of time in which studies could be made, but it proved interesting to observe how ecosystems behave in extended periods of light and darkness. When the Sun is above the horizon twenty-four hours a day, Antarctica receives more heat than any other part of the Earth. However, because of the fact that more than 98 percent of the land mass in Antarctica is covered by ice and snow (an area of approximately six million square miles), 94 percent of the Sun's energy falling on its surface is re-radiated back into space as long-wave radiation. Even though temperatures in mid-summer can get above the freezing point, gale-force winds of 40 mph are present about 50 percent of the time, and wind chills lower actual temperatures. Temperature extremes can vary from almost –100 degrees Fahrenheit to 37 degrees above zero, and rarely, warmer. The ice sheet covering this dry, hostile continent is estimated to be an average of 7500 feet thick.

Cameron and company gathered valuable scientific information for air and soil temperatures, wind direction and velocity measurements, relative humidity, visible and ultraviolet light intensities, and barometric pressures for the Antarctic dry valley regions known as McKelvey, Taylor, Wheeler, Conrow, and Victoria Valleys. The measurements were collected sometimes continually, and at intervals of every three hours.

You have to be a hardy soul to carry out scientific research here. The sense of being isolated from the rest of the world is only broken with time spent on the base ham radio. Morale is maintained through the careful scheduling of work time with free time, with an emphasis on keeping busy. It is, however, a dangerous place. A slip or a fall in the wrong place can mean a plunge into an icy crevice where no rescue attempt might be possible. The threat of frostbite is always present and can happen in as little as thirty seconds of exposure time. All researchers wanting to go to Antarctica have to take a rigid U.S. Navy physical and psychological examination.

In spite of the isolation and dangers, Roy Cameron continued with his ultimate mission: to bring back 290 soil samples from 135 different sites surrounding the dry valleys. To get to the dry valleys from McMurdo, a traverse party is formed and transported to the location desired—in the case of the dry valleys, about 100 statute miles—by a caterpillar-type tread tractor adapted for snow conditions (called a Weasel, otherwise known then as the "jeep of the Antarctic"). Other vehicles used during these expeditions were small tractor-like vehicles called Sno-Cats, and in rare instances, the U.S. Navy VXE-6 helicopters. Cameron's samples were obtained at various depths within the permanently frozen ground (permafrost), with various pits dug and surface samples obtained using only carefully sterilized tools and equipment. All the samples were put in special sterilized plastic tubes, glass jars, or sacks. These samples were kept frozen between minus 25 and minus 30 degrees Celsius and sent back to the Jet Propulsion Laboratory, where Norman Horowitz and others analyzed them for microorganism content, organic materials, and other important scientific data. These samples became known as "The Cameron Samples" among researchers who would be testing the soils as possible analogs for a Martian ecosystem.

Small amounts would eventually be distributed to scientists developing biology experiments for a yet-to-be-announced spacecraft mission to Mars called *Viking*.

In 1967 Dr. Gilbert V. Levin founded Biospherics Incorporated, a health and environmental information services and products company now in Beltsville, Maryland. Biospherics evolved to provide important health information by telephone hotline to those carrying out field services in environmental assessment of Superfund sites, landfills, leaking underground tanks, aquifer contamination, public and private work space, lead in water and paint, industrial hygiene surveys, asbestos inspection, and sampling and identification through optical and electron microscopy; and to develop its own proprietary products and processes. Levin was now the President and Chairman of his own corporation. This gave him the facilities under NASA sponsorship to develop Gulliver into the Labeled Release experiment that would test the soil of Mars in 1976.

Levin proposed to NASA in 1967 that he work on the *Mariner 9* science team as a co-investigator on the Infrared Interferometer Spectrometer (IRIS) experiment, which was to make observations of Mars' atmosphere from orbit around the Red Planet. Levin was to look for any evidence of life. IRIS had been successfully used on the Nimbus B and D Earth-orbiting meteorological satellites, which permitted researchers to study the spectra of water vapor and atmospheric constituents as well as obtain temperature measurements. One could also use IRIS to look at surface rock types and gain valuable insight into the geological aspects of our planet. So why not Mars? However, *Mariner 9* was not scheduled for launch until May of 1971 and a lot would happen between 1967 and 1971, with far-reaching implications for life on Mars, contamination of Mars, and even life on Earth.

When the first of the Cameron soil samples was analyzed from Antarctica at the Jet Propulsion Laboratory in 1967, Dr. Norman Horowitz and Dr. Roy E. Cameron and others published a paper in *Science* entitled "Sterile Soil from Antarctica: Organic Analysis" in which they concluded that Antarctica supported a significantly lower number of bacteria per gram of soil than samples in temperate regions, and that some Antarctic soils are devoid of any organic material or life. Horowitz and Cameron said that in some of the soils they examined using the Allison wet-combustion method they found organic carbon but no living microorganisms. One sample tested, simply labeled "Antarctic Soil #542," seemed to have organic matter but less than one organism per gram of soil. This is scientific jargon meaning no life!

The scientists then used a Gas Chromatograph-Mass Spectrometer on soil #542 and found no evidence for organic material. Horowitz then applied the Allison method, which measures total combustible carbon in a sample, including elemental carbon. The Allison method detected carbon, but Horowitz said, "The GCMS detects true organic matter only. Soil #542 contained coal particles which were simply carbon. Hence the Allison method's result was correct but it was measuring only elemental carbon. The GCMS detected nothing in that sample, so it was correct also." Horowitz went on to claim that the studies of Antarctic soil samples were a basis on which to relax the planetary quarantine regulations imposed on all nations by the COSPAR agreement. His reasoning was based on the presumption that if most Antarctic soils did not have life, then surely Mars would not. Another study Horowitz made determined that the snow in the dry valleys sublimes (turns immediately to water vapor) to the atmosphere without melting. This would be important for an analogy to Mars because if there is no melt-water from snow, you cannot have life, so the scientist proposed. Horowitz claimed that due to a combination of dryness, lack of oxygen, and high ultraviolet light flux on Mars, the planet had very little chance of supporting the multiplication of microorganisms from Earth. This determined effort on the part of Dr. Norman Horowitz to substantially reduce the planetary sterilization techniques employed to prevent forward contamination of the planets by space probes drew heavy criticism from the American Institute of Biological Sciences Spacecraft Sterilization Advisory Committee, which reported to NASA that Horowitz should define a criterion to be met based on the microbial parameter to be permitted. Essentially what Horowitz said was that he did not "believe" life could exist on Mars. After all, he thought, if life could not be found in the soils of the dry valleys in Antarctica—which he called a paradise compared to the Mars environment—then how could life from Earth survive there? It would be a strange position for him to take considering he was developing "his" Pyrolytic Release experiment for the detection of life on Mars.

In 1970 Dr. Gilbert V. Levin was busy studying how he was going to get Gulliver to fit into a space one-fourth the size of a cubic foot, the new NASA biological laboratory constraints for the upcoming Viking Mission to Mars in 1975. Originally, another spacecraft had been proposed in 1960 to study the surface of Mars, called *Voyager* (not to be confused with the *Voyager* spacecraft that flew to Jupiter and the outer planets in the 1980s). *Voyager-Mars* was planned as a large spacecraft that would have been launched on a *Saturn V* booster rocket, the same rocket that sent men to the Moon. There was going to be a lander that would carry a host of scientific instruments including biology detection devices. Wolf Vishniac and Gilbert Levin had their experiments "The Wolf Trap" and "Gulliver" ready to fly in 1965. They would have easily fit into the *Voyager* spacecraft design. But as the program was studied more closely, it turned out that it would cost billions of dollars, so NASA decided to opt for a more cost-effective program—the Viking Program. Since the *Viking* spacecraft would be much smaller in size than *Voyager*, every single experiment had to be miniaturized to the best of abilities, or it could not fly to Mars on *Viking*. It was just that simple.

Levin began to design a new version of Gulliver that would be appropriate for the Viking Mission. Levin's method of launching projectiles with Gulliver was replaced in the new spacecraft design. There would be a robotic sampler arm on *Viking* pre-programmed to scoop up Martian soil and deliver it to a series of "funnels" located on top of the *Viking Lander* that led into the biology and other soil-based experiments, the GCMS and X-Ray Fluorescence Spectromenter (XRFS).

In 1970, Levin began the tedious search of resumés to find the right person to head up the lab work on this growing project. His search led him to a young Assistant Professor who had spent six years at the School of Hygiene and Public Health at Johns Hopkins University—Dr. Patricia Ann Straat. Dr. Straat had done general research in the areas of molecular biology and biophysics and had one quality Gil Levin could not resist—she was an expert at working with radioisotopes. Straat recalls:

> Someone from Gil Levin's office called me who heard I was looking for a job. I went down for an interview and became completely fascinated with the whole thing as soon as I heard it. I had been waiting to make a career move, and the job that Gil Levin offered me seemed to be the most exciting thing out there.

Levin taught Straat everything he knew about the radiorespirometry technique to be used on *Viking*, and they both set out to design the Labeled Release experiment as a compact module. They worked on other projects as well, and one of the first joint patents this new collaboration obtained was for a microbial radiorespirometry technique for hospitals to make a "fingerprint" of bacterial respiration to identify the bacteria for proper treatment of the patient.

Mariner 9 WAS IN ORBIT NOVEMBER 13, 1971, revealing a Mars never before encountered. As a great dust storm began to clear, the *Mariner 9* cameras revealed a dynamic atmosphere with water vapor and clouds, huge volcanoes, and areas that looked like ancient dry river beds. The southern polar cap region looked extensively eroded, as if by glaciation. The *Mariner 9* cameras took 7,329 black and white photographs. Evidence for a "wave of darkening" was considered inconclusive, because aerosols and dust in the atmosphere reduced the contrast of

RICK KLEIN/CORNELL UNIVERSITY

Mariner 9 photo taken in 1971 showing the layered atmosphere of Mars, which is comprised of almost 95 percent carbon dioxide.

fine surface detail. Dr. Carl Sagan, who worked on the *Mariner 9* Imaging Team, pointed out that a light layer of dust over darker surface materials being removed by Martian winds might be the cause of the wave of darkening as observed from Earth.

The *Mariner 9* IRIS experiment that Gil Levin worked on revealed nothing about biological processes on Mars—at least not directly. Spectroscopic measurements made by IRIS revealed that the Martian atmosphere contained about 0.15 percent molecular oxygen. Theoretically when Martian water vapor molecules are split into hydrogen and oxygen by ultraviolet light from our Sun, the molecules eventually escape into space. *Mariner 9* calculations indicated a rate of escape equivalent to 100,000 gallons of water per day. Since the molecular oxygen content of Mars' atmosphere is still the same after twenty years of observations, something must be replacing it. Oxygen liberation by biological processes such as photosynthesis was a possibility—but further testing of the soil on Mars would be

113

necessary before a definite conclusion could be drawn. [Author's note: After Viking and the hydrogen-peroxide theories proposed by many to explain the LR result, Levin queried his IRIS teammates and found that IRIS had an excellent capability to look for the spectroscopic signature of hydrogen peroxide, and none had been found.] Nonetheless, the discoveries and success of *Mariner 9* fueled the interest of the NASA biology team, because the images showed Mars once did have liquid water flowing on its surface, so life may have developed then and possibly evolved to survive. Most thought any survivors would be below the soil surface in permafrost regions.

The Russians launched two more spacecraft to Mars in 1971, with a lander and orbiter attached to each. *Mars 2* entered orbit on November 27, 1971, and *Mars 3* December 2, 1971. The lander from *Mars 2* crashed into the surface of Mars. The *Mars 3* lander survived 110 seconds on the surface and died. Eventually the *Mars 2* and *3* Orbiters would decay and crash into the planet. Again, at best, the Russians had sterilized the outside of their spacecraft, but COSPAR was never presented with any documentation of sterilization at all from them as required by COSPAR's international quarantine agreement. As far as anyone knew, Mars was being invaded by a host of Russian microorganisms. If they could survive there, and various studies demonstrated it was possible, then would *Viking* find Martian organisms or Russian-implanted Earth life?

WHEN DR. WOLF VISHNIAC READ Horowitz and Cameron's paper in *Science* concerning the sterile soils in Antarctica, he had one thought—he didn't believe it. Although he had never been to the Antarctic before, Vishniac studied what he considered the most exotic organisms—the autotrophic bacteria, which he thought capable of living in the most extreme environments. Vishniac also believed that the methods Cameron and Horowitz used to culture the Antarctic soils to look for bacteria were not reliable enough. As a key member of the Space Science Board of the National Academy of Sciences and chairman of COSPAR's biology working group, Vishniac was horrified when he learned that Norman Horowitz was leading a campaign to relax planetary quarantine and sterilization procedures implemented not only for the protection of other worlds, but Earth as well. Vishniac's knowledge of the incredible versatility of living systems on Earth demanded a second opinion on the Horowitz and Cameron conclusions. He decided he would see for himself and test the soils of Antarctica personally.

Wolf Vishniac went to see his old friend, now President of Biospherics Incorporated, Dr. Gilbert V. Levin. Vishniac said he planned a scientific expedition to Antarctica within the year and asked Levin if he would like to combine efforts to help settle the sterile soil issue and analyze the data when he returned. Levin then introduced Wolf Vishniac to his new co-experimenter at Biospherics, Dr. Patricia Ann Straat. Both Gil Levin and Patricia Straat eagerly embraced Vishniac's offer. Levin's experience within the exobiology field and his concern about forward and back contamination of the Earth and planets was well known to Vishniac. The trio of scientists sat down to discuss at length how Vishniac's scientific sojourn to Antarctica might prove what they already knew but needed to have hard data on—that almost everywhere on Earth there is a thin film of microorganisms. When Levin and Straat learned of Vishniac's old technique of implanting slides into soil as a test for colonies of microorganisms reproducing, Levin was convinced they could demonstrate beyond doubt the viability of microorganisms in the regions Cameron-Horowitz found "sterile" Antarctic soils.

It was decided that Vishniac should use a combination of three or more tests on Antarctic samples *in-situ*, a term meaning while right there in the environment. Vishniac would bring his Wolf Trap experiment, his glass slides, and a field version of Gil Levin's experiment which Dr. Straat helped him refine for the trip to the Great White Desert. Actually, the experiment that Levin showed Vishniac to use was the same technique he developed in 1951 that subsequently turned into Gulliver and later evolved as the Labeled Release experiment, only without the instrument. Levin explained to Vishniac:

> The way we run this test without an instrument [a complicated instrument] is to take a small planchet [a small flat aluminum dish about an inch in diameter with a lip around its top edge] and put a small amount of soil in it, and then put a drop of radioactive

nutrient on top of the soil. Then we take a duplicate planchet and put a getter-pad in it [a getter-pad is a piece of blotting paper, cut in a circle to just fit into the planchet so it stays there]. We then place a drop of a saturated solution of barium hydroxide on the getter-pad. This is known as the "getter" because it "gets" by reacting with any carbon dioxide released by microorganisms in the soil to form carbonate, which becomes a solid on the getter-pad. The getter planchet is removed. Then we take a second planchet cap with a barium hydroxide-moistened getter-pad in it, put it on top of the first planchet, and let it sit while organisms in the soil continue to consume the nutrient and produce radioactive carbon dioxide, which comes up out of the soil and out of the liquid. The radioactive carbon dioxide rises, hits the getter-pad, and is precipitated and changed from a gas into a solid.

The deposition of the gas on the getter lowers its concentration above the soil sample, which brings more gas up out of the soil to replace it, so you have something like a conveyor-belt of molecules rising from the soil to deposit on the surface of the getter-pad. After about fifteen minutes of incubation time, you take the getter-pad planchet off and immediately replace it with a new getter-pad planchet. You collect a series of them, perhaps ten of them at fifteen-minute intervals. Then dry the getter-pad planchets and place them into a machine that counts the radioactivity in each planchet. You just plot the radioactivity that evolved from the soil over time, accumulating them. First you have a 0 at 0, then in fifteen minutes you might have 100 counts per minute, then in thirty minutes you may have 500 counts per minute, and you build this curve on a graph or chart.

Wolf Vishniac was impressed with Levin's concept and asked how this field version of Gulliver differed from the one he was building to fly to Mars. Levin replied:

> It does the same thing the Mars instrument does automatically, except the automated instrument does not have to dry planchets. It just keeps collecting the gas in the head-space, that is, the space above the soil, and the instrument counts continually. It keeps counting and prints out at whatever interval you want, i.e., 15, 30, 45 minutes, and so on to get the same kind of plot.

The thought that Levin had developed this technique to test drinking water in 1951 and was soon to test the soil of Mars was reason enough for Vishniac to want to try it in Antarctica. Levin told him the sensitivity of the technique could detect as few as ten to fifty colony-forming cells (ten million times more sensitive for detecting living microorganisms than the Viking Gas Chromatograph-Mass Spectrometer that would also fly to Mars). Surely, thought Vishniac, "How can I possibly fail?"

In the austral summer of 1971–72, Dr. Wolf Vishniac and a student assistant from the University of Rochester, Stanley E. Mainzer, left for Antarctica to study the same dry valley regions that Dr. Roy Cameron and his colleagues studied in 1966–69. Vishniac brought his simple glass slides that he used to detect whether microorganisms were reproducing or not, something that Cameron and Horowitz claimed virtually impossible in Antarctic soil. Vishniac explained to his young assistant, Stanley Mainzer:

Dr. Wolf V. Vishniac smiles as he models his official Antarctic Research jacket for Dr. Gil Levin and Dr. Patricia Ann Straat at Biospherics Incorporated in 1973.

> When you insert the sterilized glass slides [like the kind used for microscopes] into soil and a week later pull them out, you will find whether or not microorganisms are reproducing by simply looking at the slide, because if they are not, you will not see anything. However, if you see colonies of microorganisms on the slides, it proves there are organisms in the soil and they are reproducing. (personal communication, DiGregorio and Straat, 1996)

Vishniac and Mainzer placed the slides in a variety of locations called "stations" in the Wright Valley-Asgard Range region of South

Dr. Vishniac and Dr. Straat pose for Gil Levin. Straat worked closely with Vishniac on developing a field-kit version of Levin's LR.

Victoria Land, trying to sample as many places as possible in the same areas from which Roy Cameron obtained his samples. The mythological Norse names given to the various areas seemed to fit perfectly—names like Mount Odin, Odin Valley, and the Asgard Range. In almost every case, when using the sterile glass slides in the Antarctic soil, Vishniac found microcolonies—sometimes a few hundred to several thousand colonies. He knew only actively multiplying bacteria could be responsible. Why hadn't Cameron or Horowitz tried this technique? he thought. After all, it was really just an old technique used by microbiologists for many years.

Vishniac knew, however, that if he was to prove microbial reproduction in Antarctic soils declared sterile by Horowitz and Cameron, he must be as thorough as possible and would need as much evidence as possible. Vishniac and Mainzer employed the use of the Wolf Trap light-scattering experiment and Dr. Levin's field version of his experiment. Both of these powerful instruments supported the findings that microorganisms were abundant in the Antarctic soils. They found bacteria, fungi, and algae, including diatoms. Vishniac knew the bacteria were growing because dormant cells do not stick to the sterile glass slides and generate colonies.

The most Mars-like environment on Earth—the dry valleys of Antarctica. If Vishniac could find microbes living here, perhaps they could survive on Mars as well.

Other fascinating studies Vishniac and Mainzer conducted were observations of snow melt in the dry valleys. Once again, Horowitz and others had published that the snow ice of the dry valley region sublimes directly into the atmosphere, but Vishniac saw that the snow did melt and was a major source of water for Antarctic microorganisms. It was available only intermittently, and microbes using it would have to survive long periods of desiccation. Also, the science team of Vishniac and Mainzer noticed that occasionally when brief snowfalls occurred, icicles and puddles formed and were absorbed by the soil.

From these initial studies it was deduced that microorganisms of this cold desert were moistened even less frequently than organisms in the hot arid deserts of the Earth but managed to get on just fine. The findings would constitue a major victory for those who still thought Mars could sustain life. Horowitz, confronted with the data,

119

replied that the microorganisms Vishniac found did not normally live in the soil he continued to insist was sterile, but had been dropped there by winds from the more favorable climates just before Vishniac found them!

The most interesting discovery would not be made until Vishniac and Mainzer left with their Antarctic soil samples in stow and brought them to New Zealand, where they met with a microbiologist who was an expert in electron microscopy, Dr. John Waid of the Department of Botany at the University of Canterbury in Christchurch, New Zealand.

Dr. Waid, while applying the high magnification abilities of the electron microscope, observed a curious capsule surrounding many of the Antarctic microorganism cells. He described it as "reminiscent of a honeycomb or sponge." Bacteria with such a protective capsule surrounding their cells would be able to retain water for extended periods of time as well as enjoy protection from the ultraviolet spectrum of the Sun's rays. In Antarctica, the ultraviolet ray intensity is several orders of magnitude higher than in temperate latitudes, where rays are intercepted by the ozone layer.

Another observation was that these encapsulated microbes were in communities consisting of many dead cells, as if the living cells had surrounded them and kept them stored for times when other "food" was not available—a group of scavengers. The conclusion that Vishniac and Waid came to was that the Cameron samples from Antarctica, isolated and cultured with the traditional methods employed by the Cameron science team, including those of Dr. Norman Horowitz, were unsuited to reveal existing microbial communities in the dry valley soils of Antarctica. Vishniac's work suggested that bacterial reproduction occurred at all the sites, and moreover, none of the soils were sterile. Excited with the results of his work, Vishniac made arrangements to return to the United States.

Upon his arrival in North America with data in hand, Vishniac could hardly wait to submit his scientific findings. He sent letters off to Horowitz and Cameron saying he indeed found evidence for microcolonies of bacteria in every area he tested in Antarctica. The reception Vishniac got was less than cold. He was criticized that his glass slides might have been contaminated by improper sterile technique. Vishniac fired back, saying he had employed the Wolf Trap, Levin's microbial radiorespirometry experiment, and other microbiology techniques including micrographs (pictures taken with a microscope) with an electron microscope. Vishniac submitted papers to many scientific journals including *Science*, which, mysteriously, would never publish his results.

Later, Vishniac was summoned to NASA Headquarters in Washington, D.C., where John Naugle, the Associate Administrator for Space Science and Application, had him sit down for some discouraging news. Naugle proceeded to inform Vishniac that his "Wolf Trap" experiment, originally scheduled to fly on *Viking* to Mars, had been removed from the mission to reduce costs and simplify the biological payload. Vishniac, stunned and angry, demanded to know who was responsible for removing his experiment. Naugle told him the ultimate decision was made by Joshua Lederberg, Vishniac's fellow Viking Biology Team member, and Harold Klein, the Viking Biology Team leader. The reason given was that spacecraft constraints of weight required the elimination of one of the four life-detection instruments. Since Vishniac's Wolf Trap tested for life by placing the soil in liquid water, which at the time was thought not to exist on Mars, it was the least likely of the four instruments to succeed.

If Vishniac thought this was punishment for going up against the NASA Bioscience Section (with Horowitz and Cameron in charge) with his new Antarctic data suggesting microbial life was widespread, he did not say. But it did make him more determined than ever to prove them wrong, not only about sterile soil in Antarctica but especially because of Horowitz's repeated attempts to relax the planetary quarantine laws as established by the COSPAR committee. Why would Horowitz dismiss Vishniac's data without critical review? Why would he insist that planetary quarantine laws be relaxed? What could possibly be the motive of such action? Perhaps plans were being made, pre-Viking, to bring back a soil sample? Obviously if environmental groups suspected that Mars had a microbial ecosystem, a sample returned directly to the Earth would be highly protested because of the possibility of disease-causing organisms from Mars. But if everyone thought Mars was a dead "sterile" world, then such a sample return mission could be justified in order to study the Martian soil "chemistry" and "geology." Would the military be interested in extraterrestrial organisms as a biological weapon? These questions and more would come to the forefront after the Viking search for life on Mars completed its strange mission, and today have far-reaching implications for both Earth and Mars.

When Wolf Vishniac informed Drs. Levin and Straat that his Mars life-detection experiment had been removed from the Viking biology payload, they were extremely concerned. "Vishniac was funded before me," said Gil Levin. "If anyone's experiment should fly, it should be his." However, Dr. Wolf Vishniac did not let the incident stop him from planning a second trip to the Antarctic, and this time he was going to take a geologist with him, Dr. Zeddie P. Bowen, professor of geological sciences at the University of Rochester. Vishniac told Levin and Straat he wanted to have a geologist with him to determine rock types and to drill core samples from the Antarctic permafrost. He also said that he was going to continue the sampling and collecting of soils considered sterile and demonstrate again that life does exist in them. Once more he would take his Wolf Trap, his glass slides, and the Levin/Straat field version of the LR.

A few months before his planned departure to Antarctica in 1973, Wolf Vishniac went again to take the Navy physical examination required by all who plan to visit the continent of cold. He did not pass the exam this time. The Navy personnel informed Vishniac they thought his right arm, shorter than his left from a youthful bout with polio, might pose a danger to him and the expedition. Dr. Helen Vishniac, Wolf's widow, stated in 1996, "Wolf never had a problem with the use of his right arm." So once again a stunned Wolf Vishniac returned home with thoughts of his work being slowly manipulated and destroyed. After a brief time Vishniac decided he was not going to accept the Navy's decision to keep him from going back to Antarctica. His work, he thought, was too important, both for our world and the planets of our solar system. Vishniac took the results of his earlier work with Stanley Mainzer to several politicians and wound up convincing a U.S. Senator to help him get back to

Antarctica. The unnamed politician removed the "red tape," Vishniac passed the Navy exam, and once again the microbiologist from the University of Rochester was on his way to the great white desert.

The team of Dr. Vishniac and Dr. Zeddie P. Bowen arrived at the U.S. McMurdo Station, Antarctica, in September of 1973. Vishniac's first objective was to study the water absorption of soil particles and its significance for a microbial ecology. Both Bowen and Vishniac began gathering small and medium-size rock samples that Vishniac observed had lichens

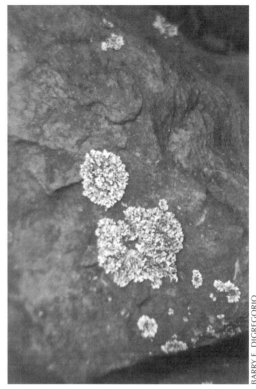

Lichens growing comfortably on a rock near Lake Ontario as temperatures hover at a cool –8 degrees Celsius.

on them which appeared to go down and into the rock. He put these peculiar samples in a sterile sample bag and the group continued work. Again, Vishniac found more of these unique rocks, a slightly translucent sandstone with lichens in the rock pores. A few drill cores were taken of the permafrost to a depth of forty centimeters. When these samples were analyzed back at the McMurdo Station biology laboratory, Vishniac observed that the numbers of microorganisms in Antarctic permafrost increased with soil depth. More profound was the fact that when he thawed the ice-cemented permafrost, almost all of the bacteria and algae were restored to vigorous activity.

Vishniac's thoughts immediately turned to Mars. If viable frozen organisms could survive in ice-cemented soils in Antarctica, maybe

123

they could on Mars too. The bacteria in these permafrost samples could have been there for ten thousand years or more and were still alive! While examining his translucent sandstone rock samples, Vishniac decided to send some to a researcher he knew at the Florida State University Department of Biological Sciences and Polar Research Center. Dr. E. Imre Friedmann would receive Vishniac's samples and make an astounding discovery. Dr. Friedmann conducted research in hostile cold regions on Earth and was known for his work in Siberia, where he took core samples from the permafrost there and found through radiocarbon dating that they were on the order of three to four million years old. The ice-cemented soil or permafrost from Siberia could be thawed out gradually and to everyone's complete astonishment, the microorganisms were still alive after three to four million years of being frozen.

Dr. E. Imre Friedmann cracked one of Vishniac's samples open and found an absolutely amazing discovery. Inside and through the rock was growing a community of organisms he called cryptobiotic organisms—lichens and algae that burrowed into the sandstone rock pores and made a home for themselves. Friedmann later officially called them cryptoendolithic [crypto = hidden, endo = in, lithic = rocks], meaning life that could live concealed inside a rock! When moisture was available and temperatures reached the melting point of water, the

DR. GILBERT V. LEVIN

Example of a rock from Antarctica containing living cryptoendolithic lichens growing down inside the rock pores. If organisms on Earth could find a home inside rocks, why couldn't they do the same on Mars?

organisms would begin metabolizing and multiplying, creating a whole community of life on the inside of a rock. This discovery could have enormous implications for any life that might exist on Mars, but not until the *Viking Lander* biology experiments in 1976 could any con-clusions be drawn. Wolf Vishniac had found these rock samples in a region of Antarctica declared dead and devoid of life. Though it would be his greatest contribution to Antarctic and exobiological studies, he would never get to see Friedmann's analysis.

Dr. Zeddie P. Bowen and Wolf Vishniac had set up their base camp near Mount Baldr in November of 1973, located within the TransAntarctic Mountain range. Vishniac wanted as many soil sam-ples as possible from the mountain slopes, because he thought if life were here on the mountains of Antarctica, the most hostile wind-blown areas on Earth, then surely life was everywhere on the Earth, just as Gil Levin and he had discussed many months ago.

On December 10, 1973, Dr. Wolf Vishniac informed Dr. Zed-die P. Bowen he was going to retrieve some glass slides he implanted in an ice field located between Mount Baldr and Mount Thor and would return to camp within twelve hours. Vishniac said if he was not back within four hours of the limit that Bowen should come look-ing for him. Vishniac followed a pre-marked trail leading him to his samples, and while picking some of them up, saw a new area he thought might be a good place to put some slides and planchets. Trag-ically, this area proved to be so slippery that Dr. Wolf Vishniac slid to the edge of a cliff in the Asgard Mountains and fell more than a thousand feet to his death.

When the four-hour safety limit passed, Dr. Zeddie Bowen set out to look for his colleague and found him at the base of Mount Baldr and Mount Thor. It would be an ironic place for Vishniac to die—in mountains named after Viking Gods—because he should have been the first to have his biology experiment flown on the *Viking* spacecraft to Mars. Zeddie Bowen hurried back to camp and radioed McMurdo Station with the sad news. A Navy UH-1N helicopter, along with members of a New Zealand mountain climbing team, recovered Dr. Vishniac's body, which was flown to New Zealand for transport back to the United States. Wolf Vishinac left behind his microbiologist wife, Helen, and their two sons, Ethan and Ephraim. Helen S. Vishniac carried on the Antarctic research her husband set out to do and later became head of the Department of Microbiology at Oklahoma State University. Helen S. Vishniac commented on Dr. Norman Horowitz's theory of the microorganisms being blown in by the wind to Antarctica, saying:

> Horowitz is wrong. Sure, some organisms are blown in as they are all over the world, but Wolf Vishniac proved indigenous life thrived in Antarctica, very specialized organisms that could not survive anywhere else but there in those hostile [to humans] conditions. Today it is commonly accepted that the only sterile soils in Antarctica are those that contain extremely high levels of toxic boron.

When Gil Levin and Pat Straat learned of their friend's death in Antarctica, they were devastated but made every attempt to save and publish Vishniac's data. Unfortunately, many of Dr. Wolf Vishniac's planchets and other materials were not labeled or identified when they were returned to the University of Rochester in New York. Whether the labels were lost in transit is still a mystery, because Vishniac was too skilled a scientist not to have documented his samples. The un-labeled samples were then sent to Levin and Straat for analysis, but without the site identification, full significance could not be attributed. The data and information salvaged from Dr. Vishniac's last trip to Antarctica were submitted by Dr. Patricia Straat to the scientific journals, but no one seemed interested enough to publish the results.

Dr. Wolf Vishniac's work pointed the way to how Mars could sustain a microbial ecosystem, but it would now be up to Dr. Gilbert V. Levin and Dr. Patricia Ann Straat to see if he was right. The next stop on this bizarre journey was Mars, where the most incredible results in the history of science would be obtained.

CHAPTER FOUR

The Good, the Bad, and the Vikings

I would like to dispel the completely unfounded notion that the Viking-GCMS experiment was designed to detect microorganisms in the surface of Mars. Instead, its purpose was to search for and identify organic compounds if any, in the Martian soil....

—DR. KLAUS BIEMANN, Principal Investigator for the Viking GCMS experiment (1996)

You have to remember two things: One is that the Viking Landers *landed in what would be more or less desert areas of Mars in order to find a very safe place on which to land, and so that kind of reduced the probability of finding organic material on the planet should it be there. Secondly, the sensitivity of that GCMS we sent there over twenty years ago is far less than the sensitivity that we are talking about here that we are applying with our Earth-based techniques....*

—DR. WESLEY T. HUNTRESS, JR., Associate Administrator for Space Science at NASA Headquarters, commenting on the Viking GCMS sensitivity at the August 7, 1996, NASA Life on Mars Press Conference

A
S THE VIKING PROJECT MANAGER, it was Dr. Gerald Soffen's responsibility to make sure the biology and GCMS
experiments were kept on schedule for their development
and integration onto the *Viking Lander*. Soffen, who began his career
as a biologist while studying at the University of California, made
several changes in agenda before coming to the Jet Propulsion Laboratory in 1961. He pursued the idea of becoming a doctor of medicine, studying both at Princeton and New York State University.
However, the thought of having a part in the search for life on other
planets was exceptionally appealing and an opportunity he could not
resist. For the next decade, Soffen would largely affect decisions on
what biology instruments would get to fly on NASA space probes.

One of the most agonizing experiences for Soffen during the
Viking program involved the GCMS experiment. At one point there
were so many problems with the device that he had the NASA projects office award another contract to a company called Bendix Aerospace Systems Division of Ann Arbor, Michigan, to study the
possibility of building an Organic Analysis Mass Spectrometer, or
OAMS, instead of the Gas Chromatograph-Mass Spectrometer
(GCMS). The problem with the Viking GCMS was that it was a
highly miniaturized version of an instrument that usually takes up
the space of a good-sized chemical laboratory. A fully automated
miniature GCMS that could be reliable *in situ* had never been achieved
in terrestrial GCMS systems.

Dr. Klaus Biemann himself, frustrated with the problems of having such a miniaturized device built to function automatically up to
250,000,000 miles from Earth, asked NASA officials that Dr. Alfred
O. C. Nier of the Viking Entry Science Team be brought in to analyze the problems. Dr. Nier was a leading authority on GCMS systems and considered by his peers to be the best in his field. Nier wrote

128

to Dr. Soffen regarding the Viking GCMS: "While I regard a properly devised and managed GCMS experiment as one of the most important things we could do on Mars, the history of this endeavor leaves so much to be desired I really wonder whether it [GCMS] has not disqualified itself already" (Ezell, 1984). Some of these problems included numerous leaks in the GCMS plumbing. If the atmosphere of either Earth or Mars entered the test cell chambers and ovens it would render any data useless. It would also be determined that any metal in the sample in combination with the primary carrier gas (hydrogen) would result in

DR. GILBERT V. LEVIN

Dr. Gerry Soffen, Project Scientist for the Viking Mission to Mars.

catalytic conversion of the organic compounds. A frequent problem that plagued the GCMS during preflight testing was the shuttle block mechanism on the surface sampler arm; this moved surface soil material from a load position to either an instrument delivery position or a dump position, but was found to be jamming on occasion. If this occurred on the surface of Mars, there would be no sample delivered to the GCMS for analysis.

Even though Dr. Klaus Biemann had made the statement many times that "the Viking GCMS is not designed to look for microorganisms," GCMS systems have in fact been used for that purpose since 1970. All GCMS instruments require that samples for study be vaporized by heating them into a gas before they can be ionized and analyzed. In the case of the Viking GCMS, which was located inside the Lander body, a sample of Mars soil is scooped up by the Viking surface sampler arm and dumped (dropped) into a funnel located on the outside of the Lander. From here, the sample enters a small grinding motor which grinds it up into a powder, and then it is delivered to one of three ovens. Once in the oven, the sample is heated quickly to 200 degrees Celsius (392°F) to release atmospheric and organic gas compounds that might be present. These gases are then pushed out of the oven and into the gas chromatograph by an on-board supply of carbon dioxide, which is labeled with an isotope (carbon-13) so it may be distinguished from the carbon dioxide in the atmosphere of Mars by the mass spectrometer portion of the instrument.

The nominal operating temperature range of the GCMS is –5 to +45 degrees Celsius. Just beyond the gas chromatograph is where the palladium-silver coil is located. Vapors from the heated samples pass through the palladium-silver coil by yet another onboard gas—hydrogen—and into the mass spectrometer, where they are analyzed. A complete mass spectrum is produced every ten seconds for one and a half hours, the time required to make a complete analysis. This same sample is then again heated to 500 degrees Celsius where any remaining organic molecules are evaporated to become a gas and again analyzed by the mass spectrometer. A third and fourth optional heating temperature of 50 and 350 degrees Celsius could also be used if the molecular science team requires additional analysis. All four temperatures—50, 200, 350, and 500 degrees Celsius—would be used on the surface of Mars.

In early chemical laboratory GCMS studies, it was routine to have a sample of dehydrated bacteria vaporized with controlled heating, and very precise measurements would be made of the molecules released, sometimes within the mass range of only 700 atomic mass units. Other techniques involved the high-temperature pyrolysis of intact bacteria, which converts chemical markers to gases with molecular weights below 150 atomic mass units. Mass spectrometry has been used for decades to characterize microorganisms using carbohydrate markers within bacterial cells such as muramic acid, D-alanine, or B hydroxyl myristic acid. A laboratory-quality GCMS can be used to differentiate whole bacterial cells or to detect microbial contaminants in complex samples as diverse as animal body fluids, tissues, and airborne dust, without the necessity of growing cultures. Bacterial carbohydrates make up a large portion of the cell wall.

Vance I. Oyama—Principal Investigator for the Viking GEx biology experiment—pioneered mass spectrometry in the early 1960s and recognized that a GCMS could be a powerful tool to look for life on Mars. Oyama became involved in the analysis of the rock and soil samples returned from the Moon, and with a large state-of-the-art GCMS in the Lunar Receiving Laboratory at the Johnson Space Center was able to detect small amounts of organic material in them. For example, from a small *Apollo 11* sample heated to 1000 degrees Celsius, about 5 parts per million of carbon material was detected with about 1 part per million believed to be from the Moon. The remaining 4 ppm were thought to come from carbonaceous meteors impacting the Moon over geologic time. Therefore, if the Moon had small amounts of organic material, scientists thought, certainly Mars must have accumulated significant amounts that could be readily measured.

One aspect of the Viking GCMS was that it employed the use of palladium-silver alloys in its hydrogen/palladium separator, something uniquely developed for this instrument alone. Once on the Martian surface there were potential problems for the palladium-silver separator (coil) located inside the GCMS between the gas chromatograph and the mass spectrometer. First, there was a possibility that any compounds in Martian soil could become reduced catalytically when the GCMS used its internal supply of hydrogen gas. If this occurred, the palladium-silver hydrogen separator might be impaired or destroyed. Also, sulfur (50–100 percent more abundant in Martian soil) and iodine compounds, if in substantial amounts, could "poison" the palladium surface of this device and inactivate the palladium. If this occurred, a possibility existed that only a part of the GCMS instrument, the mass spectrometer, might work, while the gas chromatograph would not. A study of palladium sulfur poisoning was conducted in 1974, before the launch of *Viking*, and it was discovered that when unexpected quantities of sulfur-containing gases were released during the heating phase of a GCMS sample, the palladium-silver hydrogen separator was rendered so ineffective that it had to be replaced.

Furthermore, some scientists even today do not realize that the efficiency of the GCMS in the pyrolysis mode (the 500 degrees Celsius mode) is only 10 percent. In other words, the Viking GCMS could only detect about 10 percent of the organic content of typical Earth soil. The same would be expected in the pyrolysis of Martian soil—and as much as 90 percent of the organic material could be missed.

Finally, analysis of soil and rocks returned from the Moon by *Apollo* astronaut crews demonstrated that when the lunar samples were subjected to pyrolysis temperatures of 1000 degrees Celsius or above, small amounts of organic material were liberated as a gas that could then be measured. Scientists at the Lunar Receiving Laboratory in Houston knew that for minerals to produce large quantities of gas (to measure organics in a GCMS), a pyrolysis temperature of 1000 degrees Celsius was required. The miniature Viking GCMS would not be equipped with this ability.

Ironically, the information on the Viking GCMS shortcomings would be published by a graduate student of Dr. Klaus Biemann himself. The grad student's name was John Milan Lavoie, Jr., and he submitted a Doctoral Thesis entitled "Support Experiments to the Pyrolysis/Gas Chromatographic/Mass Spectrometric Analysis of the Surface of Mars" to the Massachusetts Institute of Technology in February 1979.

Another requirement of the biology and GCMS experiments before they could be cleared and integrated into the *Viking* spacecraft were tests on the Cameron-gathered Antarctic soil samples. One of these, labeled #726, was not from the dry valley regions of Antarctica but was instead from the West Transantarctic Mountains, not far from where Dr. Wolf Vishniac took samples before his death. However, Cameron and Horowitz ran only classical bacteriological tests on these samples and were able to detect less than 1 percent of any "bugs" in the soil. Thus they concluded the soil sample (#726) was sterile like most of their other samples from this region. Antarctic soil sample #726 was sent to Dr. Gilbert V. Levin and Dr. Patricia Straat, representing the Biospherics Incorporated Labeled Release experiment, and to Dr. Klaus Biemann and the GCMS molecular analysis team.

Biemann's team tested Cameron's Antarctic soil #726 in the Viking GCMS Test Standards Model and found no evidence for organic material in it, even though when they tested the sample by the Allison method, they found 300 parts per million of organic matter—a fleck of anthracite coal dust, they reported. Levin and Straat tested Cameron's Antarctic soil #726 prior to the launch of *Viking* as well. However, their results with the Viking Labeled Release Test Standards Model found living microbes in the soil. What did it mean? It means that the Labeled Release experiment, being ten million times more sensitive to living organic material than the Viking GCMS was to total organic matter (living and dead), could detect life in soils with too little organic matter to be detected by the GCMS. The microbes in Antarctic soil #726 were well below the sensitivity limit of the Viking GCMS. If Martian organisms were in the soil, the GCMS would give a negative result unless there was a large accumulation of dead organic matter. To put it more dramatically, if the *Viking Lander* with the GCMS had landed in California instead of on Mars and tested Death Valley or Mojave Desert soil, which contains approximately 1,000 to 100,000 cells per gram of soil, then the results from the GCMS data would record that Mojave Desert soil is devoid of organics and therefore sterile. What would happen on Mars with such an instrument?

In their preparations to get the various experiments ready for the flight to Mars, Levin and Straat documented their test on Antarctic soil #726 as active biology and filed it away, unaware at that time of the negative response Klaus Biemann had obtained for organics in Antarctic soil #726. Little did they realize that later their test would provide an important clue to the sensitivity of the Viking GCMS experiment on Mars, and would ultimately end in a clash with Biemann and some of the other NASA Viking scientists who refused to consider the Labeled Release data as reliable evidence against the GCMS. Dr. Harold Klein himself, the leader of the Viking Biology Team, would later accuse Levin and Straat of having conducted the test on Antarctic soil #726 after the Viking Biology Mission concluded, when the sample had perhaps become contaminated with microorganisms (more on this later, in Chapter Five).

AFTER BEING SUBJECTED TO A large number of acceptance test checkouts such as operational systems, thermal tests, and vibration tests,

The Viking Biology Instrument Package (containing all three biology experiments: GEx, LR and PR) before placement inside the Viking *Lander. It weighed 33 pounds and was contained in a space less than a cubic foot.*

the biology instrument package and GCMS were placed in a Mars environment chamber consisting of 95 percent carbon dioxide with a temperature range of –18 to 30 degrees Celsius. Then it was sterilized in a nitrogen gas atmosphere at 120 degrees Celsius for 54 hours. Upon being integrated into the *Viking* spacecraft prior to launch, the entire lander would be sterilized at the Kennedy Space Center Sterilization Chamber in an atmosphere of nitrogen at temperatures of 116 degrees Celsius for over forty hours. The GCMS in *Viking 1* was retested after sterilization and found to have a leak in the mass spectrometer vacuum envelope. Engineers decided to launch the *Viking 1* GCMS with the leak because they thought it would not worsen. The twin *Viking* spacecraft were then mated to their Titan IIIE-Centaur booster rockets, first *Viking 1* and a month later *Viking 2*, and readied for their separate launches from Pad 41 at the Kennedy Space Center.

Dr. Gilbert V. Levin and Dr. Patricia Straat were at the Kennedy Space Center launch complex on August 11 to witness the lift-off of the Mars-bound *Viking 1* spacecraft, but due to a malfunction in a thrust vector control valve, the launch was scrubbed. Another launch attempt was made on August 13, but again was scrubbed because the *Viking 1 Orbiter* batteries were showing a reading far below what they should be. Since the process of replacing the batteries on the *Viking Orbiter* is long and tedious, the *Viking 1 Orbiter* and *Lander* package was removed from its Titan-Centaur booster rocket while on the launch pad and replaced with the second *Viking Orbiter* and *Lander.* Levin and Straat, having other pressing business matters, returned to Biospherics Incorporated in Maryland after patiently waiting while NASA hurried to resume a launch schedule, but they could not wait long enough. The *Viking 1* spacecraft blasted off for Mars at 5:22 P.M. EDT on August 20, 1975, with *Viking 2* following a few weeks later at 2:39 P.M. EDT on September 9. Levin and Straat, with a number of other Viking Science Team members, watched these historic lift-offs from the convenience of the NASA Headquarters television monitors. Next stop—Mars!

Once in space and with solar panels deployed, both spacecraft located the star Canopus for use as a navigational beacon, and the second-stage Centaur rocket sent the spacecraft hurtling towards Mars, 200 million miles distant. During the next ten months in space—the time it took for *Viking* to reach Mars—there would be several instrument checkouts in-flight that would have special consequences on the surface of Mars. On the cruise to Mars, the twin *Viking* spacecraft would experience the deep cold of space, –453 degrees Fahrenheit.

The Viking spacecraft were internally powered by a small radioisotope (nuclear) generator. On the cruise, a small lamp inside the *Viking Lander* was used to calibrate the Lander cameras for color imaging so that when *Viking* touched down on the surface of Mars, it could send back an accurate interpretation. In fact, a total of fourteen internal calibrations were made between the sterilization process and landing on Mars. Twelve of these were done prior to launch, one during the voyage, and another just before the separation of Orbiter and Lander. A color test reference chart mounted on the *Viking Lander* was used to calibrate the cameras. Once this was accomplished, other verification tests could be performed. This was done because it had been discovered on previous spacecraft missions (such as *Mariner 9*) that cameras, once out in space, do not have the same response they had in the laboratory. Aside from the obvious problem of radiation, imaging scientists are not quite sure why this happens. So they rely on in-flight calibration.

On a test of the Viking GCMS experiment while in the cruise phase to Mars, it was revealed that one of the miniature ovens on each lander was not functional. This was not looked upon as critical because the GCMS had three ovens in which to heat soil samples. In a test of the ovens that were working, scientists in the JPL control room got a reading that something organic was in the test chambers! It turned out that miniscule amounts of cleaning fluids such as benzene, methyl chloride, and freon-E and other chlorinated compounds seemed to be in the instrument. Klaus Biemann thought this was a wonderful glitch because now he could measure the contaminants to prove his instrument was working perfectly. Indeed, both Viking GCMS instruments recorded that these cleaning fluids were aboard.

The Ultimate Touchdown

Five days before arriving at Mars, the *Viking Orbiters* obtained video images of the planet and one of its small moons, Deimos, for the purpose of navigating the Mars Orbit Insertion engine burn. This maneuver would place each spacecraft in orbit around Mars and could later be altered to bring the *Viking Orbiters* in an even closer orbit around the planet. The Mars Orbit Insertion rocket engine burn lasted forty minutes.

Viking 1 was in orbit June 21, 1976, and after several weeks of deciding upon an appropriate landing site, the Chryse Planitia region of Mars in the northern hemisphere was finally chosen. It looked like a safe, flat plain to land on and a good one for life-seeking because of ancient dried river channels that formed the area. It was believed Chryse might contain outwash deposits from the river channels and possibly a source of groundwater. Of course, this meant that the chance of finding evidence of past or present life would be enhanced. Although landing in the dark regions of Mars would have been best

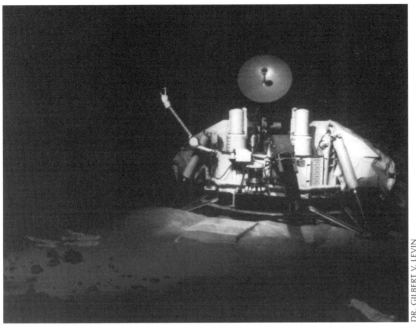

DR. GILBERT V. LEVIN

The Viking Science Test Lander at the JPL facility, used for simulations of actual sampling events on Mars.

for seeking life and organic molecules, it was just too dangerous because of the rough terrain. At last—the search for life on Mars was about to commence.

At the pre-selected orbital point, the *Viking* spacecraft computer gave the command for the Orbiter and Lander to separate. A short time later, the Lander deorbit burn engines were fired and the *Viking 1 Lander* coasted slowly downward towards the surface of Mars for a period of about 2 hours and 25 minutes, when it began to experience drag on its protective aeroshell from the Martian atmosphere, slowing the spacecraft. As the Martian gravity pulled the spacecraft down, a parachute from the Viking aeroshell was deployed at 19,500 feet. Then the aeroshell separated from the Lander and three landing legs unfolded and locked into position. At 4,600 feet from the surface of Mars, the terminal descent engines fired to slow the spacecraft, and the parachute was released. The spacecraft was now coming down at about 8 feet per second, and within minutes, kicking up a cloud of dust and debris, touched down on the surface on a very bright, sunny Martian afternoon on July 20, 1976. (*Viking Lander 2* would land in the morning on September 3, 1976.) We were on Mars at long last and about to find some of the most unusual scientific discoveries in all of recorded history.

Upon the craft's touchdown on Mars, an enormous burst of enthusiastic cheers broke out in the Jet Propulsion Laboratory in Pasadena, California, the control center for the *Viking* spacecraft. Every technician, team leader, and scientist was shouting and clapping. Some began to cry happily. It was an amazing moment of triumph for the human race. Everyone anxiously awaited the first black and white image to come in on the JPL monitors. It was SOL 1 for the *Viking Lander 1*, meaning Mars day one, and the *Viking 1* number-one camera began to transmit the first photograph, a view looking down at the footpad of the *Viking Lander*. Then within minutes a second *Viking 1* camera began to take a 300-degree panoramic view of the new landscape in black and white. As these images came in on the JPL television monitors, the crowd of scientists, technicians, team leaders, and media personnel again cheered, clapped, and cried. It was an unbelievable sight. The pictures were razor-sharp. The Martian sky appeared clear, and all around the *Viking Lander 1* were rocks and boulders. It was amazing the lander did not impact a boulder, everyone commented.

It was an emotional day that July 20th, and Gil Levin and his son Ron left JPL late to return to the apartment they had rented in Pasadena for the next three months while Dr. Levin oversaw *Viking* perform the LR biology tests on the soil of Mars. Dr. Patricia Ann Straat also rented a nearby apartment to monitor the experiment results from Mars. Gilbert and Ron Levin looked forward to the next day, July 21, because they knew the first color images from Mars would be coming in. Levin's son Ron had just graduated from high school and was eagerly anticipating spending time with his father at JPL and witnessing history being made. Little did father and son realize that come July 21, their whole perspective on how NASA conducted its affairs on another world would change, a foreshadowing of things to come. Gil Levin knew the first biology samples were not scheduled to be taken until SOL 8, or the eighth day on Mars, and so it was a time for him and his son to relax and enjoy watching the images come in, or so they thought!

It would not be Dr. Levin or any of the other biology team scientists who would command the surface sampler arm of the *Viking* spacecraft to scoop up soil and put it in the different experiment funnels. Nor would the Viking Biology Team command the biology experiments on Mars. This duty was left to the members of the Science Test Lander and Surface Sampler Team, headed by Dr. Donald Pike. These teams had to initiate computer command sequences to the *Viking Orbiter*, which were downlinked to the *Viking Lander.*

The surface sampler arm activities, along with all the other experiments on the *Viking Lander*, had to be pre-programmed. The Biology Team, however, could suggest to the Surface Sampler Team where the surface sampler should "dig" for a sample if other areas around the Lander looked biologically interesting. If a problem were to develop, the surface sampler was designed to shut down or give a "no-go" until the Surface Sampler Team could figure out what was causing the problem. To help them solve problems, the Surface Sampler Team had a full-scale working *Viking Lander* located in the Jet Propulsion Laboratory called the Science Test Lander. This was used to do exact reproduction of the surface sampler activity on Mars. The Science Test Lander was put in an artificial landscape to simulate the exact placement and size of the rocks surrounding the *Viking Lander.* They constructed this artificial Mars environment by analyzing the images returned from the surface by the real *Viking Lander.* In this way, any problems with "steering" the sampler arm could be avoided (i.e., hitting rocks instead of soil).

July 21, 1976

At about 2:00 P.M. PDT, the first color image from the surface of another planet, Mars, began to emerge on the JPL color video monitors located in many of the surrounding buildings, specifically set up for JPL employees and media personnel to view the *Viking* images.

Gil and Ron Levin sat in the main control room where dozens of video monitors and anxious technicians waited to see this historic first color picture. As the image developed on the monitors, the crowd of scientists, technicians, and media reacted enthusiastically to a scene that would be absolutely unforgettable—Mars in color. The image showed an Arizona-like landscape: blue sky, brownish-red desert soil, and gray rocks with greenish splotches. As with the previous pre-programmed black and white imaging—with panoramas taken in front and back of the Lander—so too was the color imaging sequence programmed for a panorama in front of the Lander and behind it. However, because the image was so spellbinding, only the first color image remained on the monitors for this day. Gil Levin commented to Patricia Straat and his son Ron, "Look at that image! It looks like Arizona." *(See Color Plate 5.)*

Two hours after the first color image appeared on the monitors, a technician abruptly changed the image from the light-blue sky and Arizona-like landscape to a uniform orange-red sky and landscape. *(See Color Plates 6, 7, and 8.)* Ron Levin looked in disbelief as the technician went from monitor to monitor making the change. Minutes later, Ron followed him, resetting the colors to their original appearance. Levin and Straat were interrupted when they heard someone being chastised. It was Ron Levin being chewed out by the Viking Project Director himself, James S. Martin, Jr. Gil Levin went immediately and asked, "What is going on?" Martin had caught Ron changing all the color monitors back to their original settings. He warned Ron that if he tried something like that again, he'd be thrown out of JPL for good. The Director then asked a TWR engineer assisting the Biology Team, Ron Gilje, to follow Ron Levin around to every color monitor and change it back to the red landscape.

What Gil Levin, Ron, and Patricia Straat did not know (even to this writing) is that the order to change the colors came directly from the NASA Administrator himself, Dr. James Fletcher. Months later, Gil Levin sought out the JPL Viking Imaging System technician who actually made the changes and asked why it had been done. The technician responded that he had instructions from the Viking Imaging Team that the Mars sky and landscape should be red and went around

DR. GILBERT V. LEVIN

Ron Levin (center) sits between Norman Horowitz (left) and Klaus Biemann (right) after being chastised for changing the JPL color monitors back to the original blue sky settings.

to all the monitors "tweaking" them to make it so. Gil Levin said, "The new settings showed the American flag [painted on the Lander] as having purple stripes. The technician said that the Mars atmosphere made the flag appear that way." Levin scientifically argued, "If atmospheric dust were scattering red light and not blue, the sky would appear red, but since the red would be at least partially removed by the time the light hit the surface, its reflection from the surface would make the surface appear more blue than red. There would be less red light left to reflect. And what about the sharp shadows of the rocks in the black and white images yesterday? If significant scattering of the light on Mars occurred, the sharp shadows in those images would not be present, or at best, would appear fuzzy because of diffusion by the scattering!" Levin pointed out that any student in freshman physics learns about Rayleigh scattering, which causes any atmosphere of colorless gases to produce a blue sky when sunlight passes through it.

Jurrie J. Van der Woude and Ron J. Wichelman worked on the Viking Image Formatting and Processing section of JPL. Mr. Van der Woude recalls:

Both Ron Wichelman and I were responsible for the color quality control of the *Viking Lander* photographs, and Dr. Thomas Mutch, the Viking Imaging Team leader, told us that he got a call from the NASA Administrator asking that we destroy the Mars blue sky negative created from the digital data.

Also, NASA did not want to release a photograph to the press depicting the American flag on the RTG tank of the *Viking Lander* with deep purple and yellow stripes. Try getting that on the cover of *Life* magazine! Half of the Mid-west would be taking their pitchforks to JPL, saying, "What the hell did you do to our flag?" So, we had a photograph that showed the flag red, white, and blue. We had to tweak the colors until we had an American flag again. (personal communication)

While this explanation sounds fine by itself, the first color photograph from Mars did not have the American flag in it at all! The photograph Mr. Van der Woude refers to is the second pre-programmed panorama, which would not be released to the media—a view looking back over the Lander with the American flag on the Viking RTG (Radioisotope Thermoelectric Generator) wind screen. This view would be seen many times during both Lander missions, but not with a blue sky.

It would later be claimed that the *Viking Lander* camera system could not render the true colors of Mars because of possible radiation damage to the infrared diode filters in the camera, which the imaging team said was necessary for reproducing the "true" colors of Mars. All of the multispectral imaging (color photographs) taken from spacecraft use small electronic color diodes (comparable to filter wheels) with three primary colors—blue, green, and red—and in the case of *Viking*, three near-infrared diodes (IR1, IR2, and IR3, not needed for color imaging) to construct a very accurate color image. Three black and white images are made of the same view, with each of the three primary color diodes to make one color image.

Recall the spectacular images sent back by the *Voyager* spacecraft from outer planets Jupiter and Saturn. These spacecraft color imaging systems had no problem obtaining excellent color images using the same exact technique. The Hubble Space Telescope has no difficulty imaging planets, galaxies, and nebulae. In fact, a series of February 1995 images of Mars taken with Hubble shows vast greenish areas (which some scientists still claim is an illusion due to the contrast of the orange desert regions and "gray" dark regions! Someone better tell the Hubble Space Telescope it's an illusion!), wispy white clouds, and a bluish atmosphere on the limb of the planet. Again, the images are produced by scanning an object (in this case Mars) three times using three separate color diodes for each exposure (green, blue, and red filters) and then combining all three to make one photograph.

A near-infrared diode is not necessary to reproduce color images from space, so why did the NASA Viking Imaging Team feel it was necessary on Mars? What would falsely coloring the surface of Mars in a uniform orange-red color accomplish anyway? It did help to make Mars seem unearthly! Did NASA suddenly decide that it was too soon for human society to try and accept that life on another world might look much like Earth? The story only gets more interesting.

The NASA Administrator may have ordered the blue sky "negative" destroyed, but the second color panorama (also a blue sky) looking behind the *Viking 1 Lander* and showing the American flag was, ironically, reproduced in a NASA publication. This remarkable photo was included in the "Viking-Mars Exhibit Package, stock #033-000-00711-6, NASA.43 V69-INFO." *(See Color Plate 9.)* This package contained eleven full-color posters ranging in size from 12x16 to 30x40. The caption to the second blue-sky image says: "This view is over the back of *Viking Lander 1* in Chryse Planitia. *(See Color Plate 10.)* The region is generally boulder-strewn—broken by rock outcroppings, larger boulders, and drifts of fine, crusted dust. Much dust is suspended in the atmosphere—scattering light to fill shadows and produce a sky coloration that is generally gold [Author's note: it's gold and blue] with perhaps a tinge of pink." This package also had two other views of Mars that depict green spots on rocks, even though the color had been changed to show a reddish-orange landscape. One of these shows an image of a boulder 25 feet away from *Viking Lander 1* called "Big Joe," which is about 40 inches high and 7 feet long.

(See Color Plate 11.) Big Joe would be imaged over a thousand times, and depending on what season it was on Mars, the green areas would either be prominent or disappear. [Author's note: keep in mind that Gil Levin would not see this image until I sent it to him in 1994, nor was he aware of the *Viking Lander 1* second panorama depicting a blue sky—read on!]

Back at JPL, the Viking Imaging Team released the new orange-red "first color image" from Mars with the following statement:

> This color image of Mars was taken July 21—the day following *Viking 1's* successful landing on the planet. The local time on Mars is approximately noon. The view is southeast from the *Viking.* Orange-red surface materials cover most of the surface, apparently forming a thin veneer over darker bedrock exposed in patches, as in the lower right. The reddish surface materials may be limonite (hydrated ferric oxide). Such weathering products form on Earth in the presence of water and an oxidizing atmosphere. The scene was scanned three times by the spacecraft's camera #2, through a different color diode each time. To assist in balancing the colors, a second picture was taken of a test chart mounted on the rear of the spacecraft. Color data for these patches were adjusted until the patches were an appropriate color of gray. The same calibration was then used for the entire scene. Another more pink, gray, and blue version [the first blue sky photo] was released earlier... This interpretation has been modified with further processing.

And with that statement issued, NASA released the new version to the press, photo designation: VIKING 1-54 P-17164 (color), SOL 1. The press went wild with the new image, with many headlines reading to the effect "Mars sky turns from blue to red," but not as red as some of the faces of those on the Imaging Team. The day after the color picture caper, Ron Levin walked by the outer door of the Viking Imaging Team room and saw a sheet of paper with the following poem written on it:

> Red sky at night, Martian's delight, Blue sky at two, just will not do, Pink sky at noon, this must be Brigadoon!

Under the poem someone had hand-scrawled: "Boy, you guys have a lot of guts."

By July 27, the Viking Imaging Team had perfected the process of "tweaking" the colors of Mars. Another image from the *Viking Lander 1* designated VIKING 1-57 P-17173 taken on SOL 6 showed an orange-red landscape and sky but only this time with an American flag that had the correct colors—red, white, and blue! The offical press release reads:

> The flag of the United States with the rocky Martian surface in the background is seen in this color picture taken on the sixth day of *Viking Lander 1* on Mars (July 26th) ... to the right is the Reference Test Chart used for color balancing of the color images. At the bottom is the GCMS Processor Distribution Assembly with the wind screens unfurled, demonstrating that the GCMS cover was deployed properly. . . .

While the colors of Mars were an interesting and important issue, Dr. Gil Levin and Dr. Pat Straat knew that more pressing matters lay ahead—the Viking Labeled Release biology experiment and the search for life on and just under the soil surface of Mars.

Sol 8—Viking Digs In (July 28, 1976)

The Biology Team, Inorganic Chemistry Team, and Organic Analysis Team selected surface sampler arm digging sites based on the photographs coming in on the JPL monitors and images subsequently printed out. An area two to four meters distant from the Lander at an azimuth of about 120 degrees was the limit for surface sampler diggings. As each of the science teams picked a digging site, they relayed it to the Surface Sampler Team. The Surface Sampler Team wrote the computer commands necessary to carry out the dig. The decision where to tell the Surface Sampler Team to dig was planned sixteen days before a science experiment command, with five or six days elapsing after the command before the data could be analyzed.

Since *Viking* was designed primarily as the "search for life," the Biology Team had first preference for a soil sample. Ordinary names were given to the trenches dug by the surface sampler arm. The historic first digging area on the Chryse site was given the name "Sandy Flats" (even though no sand would ever be found at either Viking landing site). The sampler arm dug into the Martian soil to a depth of about 3.5 centimeters. The process took about two hours and delivered the first scoop of soil to the three biology experiments, which shared a common funnel on the *Viking Lander*. After the soil went into the funnel, it went down into an area where it was ground up and then delivered in whatever amount the test chambers in each biology experiment could hold. The Pyrolytic Release experiment test cell held 0.25 cc of soil, the Labeled Release experiment held 0.5 cc of soil, and the Gas Exchange experiment held 1.0 cc. Each experiment could now run its individual test and await the results, which would come in at different times depending on how long the sample was incubated.

After the Biology Team had acquired its samples, the surface sampler arm went back to Sandy Flats to scoop up another soil sample for the GCMS. Dr. Klaus Biemann anxiously awaited the delivery of the sample. A monitor in the control room of the Surface Sampler Team was to display a number when the GCMS had a sample in its test oven. It never happened. Something had gone wrong.

Dr. Klaus Biemann (center—standing) with on-lookers at JPL, shortly after the Viking 1 *landing.*

DR. GILBERT V. LEVIN

Images of the surface sampler arm showed the arm over the funnel of the GCMS, but Biemann could not get an indication of a sample having been delivered. Since the GCMS had already lost two of its test ovens, Biemann decided to wait for another dig on SOL 14 rather than risk wasting an oven on a sample that may or may not be there.

The Inorganic Chemistry Team also experienced a delay on SOL 8. Apparently the surface sampler arm gave its first "no-go," a term used by the Surface Sampler Team to indicate that the surface sampler arm computer commands were aborted. The problem, as later found, was caused by the failure of a rotational switch. The Viking Surface Sampler Team went to work immediately on this problem. Biemann's instrument was without a sample, and now even the X-Ray Fluorescence Spectrometer (inorganic analysis experiment), or XRFS, would not get any Martian soil until SOL 29.

July 30, 1976

Dr. Levin and Dr. Straat had injected their Martian soil sample with radioactive nutrients on SOL 8 (July 28). The Mars atmosphere inside the LR test cell was adjusted with helium to a total pressure of 65 millibars to prevent the liquid nutrient from boiling under the atmospheric pressure of Mars, should it be 6.0 millibars or lower. Immediately, the radiation counter inside the LR started to register radioactive gas evolving from the soil. The counts per minute or cpm soon reached 2,215. Because of background radiation and a minor residual from purging the radioactive nutrient, the baseline used to measure the gas evolved from the soil was 500 cpm. The radioactivity was measured periodically every 16 minutes. By early evening on July 30, it was obvious that a significantly positive reaction was occurring. It looked very similar to the response from Earth soils containing microorganisms.

As the LR count rate continued, Gil Levin decided to have someone go out and get a bottle of champagne. While the all-important control experiment had not yet been run, it was perhaps the most critical defining moment in the history of science—life on another

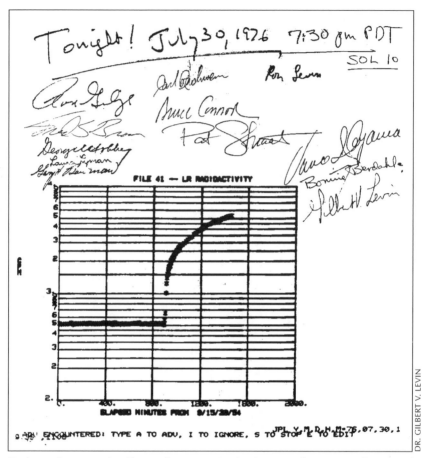

Perhaps the most significant document in the history of science—the signed LR data strip indicating the presence of life on Mars on July 30, 1976.

planet, discovered for the first time. At approximately 7:30 P.M. PDT Levin took the data page graph from the JPL control monitor and passed it around for everyone in the room to sign. He captioned the sheet "Tonight!" after the then-popular *Westside Story* song "It All Began Tonight!" Everyone in the room shared a common look of disbelief but all realized why we came to Mars—to look for life. Some of the people in the room who signed the data strip were Dr. Patricia Ann Straat, Ron Levin, Vance Oyama (GEX experimenter), Ron Gilje (Biology Team assistant), Bonnie Bendahl, Dr. George Hobby (co-experimenter with Dr. Norman Horowitz on the Viking PR experiment), and Dr. Gilbert Levin. In all, twelve people would sign the historic piece of paper. However, things have never been easy for Gil Levin, and his discovery of life on Mars from this moment on would be challenged relentlessly.

Before Levin and Straat could be absolutely sure that it was life giving their LR instrument a positive response on Mars, a control was built into the LR experiment to make sure life and not chemistry provoked the response. The Viking Biology Team decided to do this for all the Viking biology experiments: to heat a duplicate sample at 160 degrees Celsius for three hours. This had proven adequate to kill microorganisms in Earth soils. Any chemical responsible for the reaction would likely survive this heating. After heating the soil, allowing it to cool, and awaiting execution of the LR experiment again, Levin and Straat examined the data strip. Whatever it was in the Mars

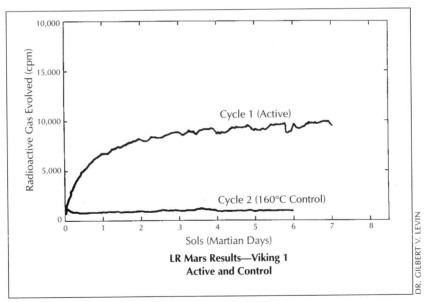

Graph depicting the first LR test conducted on Mars by Viking 1. Cycle 1 shows a steep rise in radioactive gas after the addition of Levin and Straat's nutrient injection. Cycle 2 has a flat response after Levin and Straat raised the temperature in the LR chamber to 160 degrees Celsius. Were Martian microbes destroyed by this heat or was it exotic chemistry, as NASA began to argue?

149

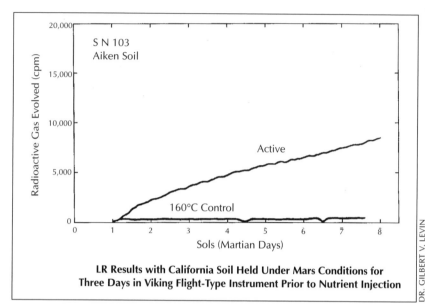

S N 103
Aiken Soil

Active

160°C Control

Radioactive Gas Evolved (cpm)

Sols (Martian Days)

**LR Results with California Soil Held Under Mars Conditions for
Three Days in Viking Flight-Type Instrument Prior to Nutrient Injection**

DR. GILBERT V. LEVIN

*Graph showing the LR results from a California soil kept under Martian con-
ditions. The reaction of the soil is almost the same as the result obtained from
Mars on the graph shown on page 149. Even the raise in temperature to 160
degrees Celsius, killing the California microbes, was almost identical to that
observed on Mars.*

soil that caused the initial reaction was now completely inactivated—
or dead. This major hurdle had been passed. Levin and his co-experimenter proved two things about Martian soil: something in the

soil was "eating" the
radioactive nutrients
and then releasing
radioactive gas. Plus,
whatever was doing it
was destroyed by heat
sterilization. The LR
experiment met all
Viking pre-mission
criteria accepted by
the scientific commu-
nity for the discovery
of life on Mars.

DR. GILBERT V. LEVIN

*Dr. Patricia Ann Straat explains the LR results
obtained by* Viking 1 *to media and JPL personnel.*

150

IN ALL, LEVIN AND STRAAT would conduct nine LR experiments between both *Viking Lander 1* and *2* covering a time period of 195 days on Mars. The season in the northern hemisphere would change from summer to autumn before the biology tests concluded.

The Controversy Begins

[Author's note: Since I have already described the way the PR (Pyrolytic Release) and GEx (Gas Exchange) experiments work in the Introduction of this book, please refer back to that information if necessary.]

VANCE OYAMA AND HIS GEx experiment were getting their share of action from the surface of Mars. On SOL 9 and over a three-day period, a 1.0-cc sample of Martian soil was incubated after Oyama had his instrument supply a 0.57-cc water-nutrient vapor along with some of Mars' atmosphere. The GEx was designed to have three operating choices: the nutrient could be used to humidify the soil as a vapor; it could moisten the soil sample with liquid nutrient; or the test chamber could analyze the soil without any nutrient vapor or moisture. For this occasion, the soil sample was contacted with nutrient vapor only and in the dark with a controlled temperature range of 8 to 10 degrees Celsius. Essentially, the GEx monitored compositional changes

DR. GILBERT V. LEVIN

Vance Oyama (did not have a Ph.D.!), Principal Investigator of the Gas Exchange Experiment (GEx), in JPL lounge in 1976.

in the atmosphere of the test chamber (above the soil) by means of a gas chromatograph.

What happened next was very surprising. Carbon dioxide from the Martian atmosphere inside the test chamber started to be absorbed by the soil sample! Also, a small amount of oxygen burst out of the soil, peaking and leveling off by the end of the third day. At first it appeared that photosynthesis of some kind was occurring because carbon dioxide was being absorbed at the same time oxygen was given off—the same reaction photosynthetic organisms on Earth have. However, since the experiment was carried out in the dark, oxygen release by photosynthetic organisms did not seem likely, unless Martian organisms had some way to extract oxygen from the soil. [Author's note: Many facultative anaerobic organisms do just that! Why it was not considered is still a mystery.] The amounts of oxygen evolved, though very small (1–10 parts per million), were fifteen times higher than scientists could account for by physical absorption alone. However, it had been known since 1951 that hydrogen peroxide is produced internally in light or dark reactions by phytoplankton on Earth (Mehler 1951). True, any oceans on the surface of Mars have long vanished. But what if such an organism evolved in a shallow Martian lake or ocean, and then continued to adapt and evolve to moist-soil conditions? Since 1965, soil scientists have been aware of the production of extracellular hydrogen peroxide by cyanobacteria (Van Baalen, 1965). Complex enzyme reactions in cells using ascorbate peroxidase and catalase for scavenging hydrogen peroxide is a controversial subject among microbiologists today. Any organism capable of scavenging hydrogen peroxide from its environment or purposefully excreting it into the soil would always have a source of oxygen and water. Could such a survival mechanism have been in place before the biological production of oxygen in Earth's atmosphere 2.5 billion years ago? Why didn't Oyama and other NASA exobiologists consider this at the time of Viking? Immediately after reviewing this peculiar data from the GEx experiment, Oyama suspected a chemical (carbon suboxide) in the Martian soil that liberated oxygen upon humidification with vapor.

On SOL 16, Oyama decided to see what would happen next to this same soil sample that had been incubated for eight days. This time he chose to inject 2.27 cc of the liquid nutrient he called "chicken soup" on top of the soil sample. If chemicals were responsible for liberating oxygen when humidified, then certainly the addition of liquid onto this soil should liberate even more, Oyama's GEx Team speculated. Once wet, the soil sample again began to absorb carbon dioxide from the Martian atmosphere inside the test chamber, but oxygen, unlike the initial reaction on SOL 9, actually decreased! At this point a biological interpretation was still open. Oyama was not convinced and knew that if he heated a sample at 145 degrees Celsius, superoxides or metal peroxides would survive because they would only decompose at 425 degrees Celsius or above, whereas living organisms would be destroyed or significantly reduced.

Oyama obtained a second soil sample from Sandy Flats and then heated it to 145 degrees Celsius for 3.5 hours to release any moisture that might be in the soil. Once the sample was heated, Oyama applied humidity, and within 2.5 hours measured the oxygen level with the GEx gas chromatograph and found only 48 percent of the oxygen originally given off by the first sample (Oyama, 1977). Oyama concluded from this and his other experiments on Mars that oxygen was present as peroxides or suboxide.

Unlike the LR experiment, the GEx emitted oxygen a second time. This was something quite different from what Levin and Straat found in their LR sample. After all, the heating of the LR to 160 degrees Celsius completely stopped their reaction. Because Oyama thought living organisms could not survive 145 degrees Celsius heat for 3.5 hours, he would later comment that carbon suboxide polymers "had all the characteristics to simulate the results observed in the Viking PR experiment," Dr. Norman Horowitz's life-detection device. Perhaps Oyama's most interesting theory was that Martian soil contained a variety of carbon suboxide polymers, each with a different reflectance spectrum that absorbed in the infrared. This, he thought, could be a global mechanism for the wave of darkening. As moisture from the polar ice cap is released, carbon suboxide would undergo striking color changes, as seen for a hundred years by Earth-based astronomers. Could purple and blue soil indicate carbon suboxide in the soil on Mars?

As soon as other scientists and researchers heard Vance Oyama's theories and comments on carbon suboxide and peroxide in the soil of Mars the stage was set for a scientific feeding frenzy that would dissuade many from pursuing a biological explanation. Dr. Levin comments on the GEx:

> In obtaining results from the GEx experiment, it is necessary to make assumptions about the absorptivity and other factors of the soil that is being tested. Thus there are two constants that have to be used as multipliers to determine the amounts of each gas evolved from the experiment. They did not have any Mars soil that they could measure for those two parameters. So those parameters were presumed on the basis of what they thought Mars soil was like. Therefore the results published for the GEx— the pulse of oxygen—are based on the numbers that were assumed for those parameters in the soil of Mars.

The GEx reaction in the presence of water vapor alone made Levin think it was not peroxide, which would not react with water vapor. Peroxide merely dissolves in water, it doesn't react with water, but nobody paid any attention to that.

One of the most puzzling aspects of the GEx experiment that went virtually unnoticed prior to the Viking Mission was its unique ability to give a positive reaction for life using sterile lunar soil returned by the Apollo astronauts! As in the case of the Antarctic test soils, the Viking Biology Team and GCMS Team received samples of lunar soil with which to test their instrument's sensitivity. In one test of the GEx with a lunar soil, a decline in oxygen and carbon dioxide with a sudden rise in hydrogen indicated changes that the GEx would consider life! In a test of the LR by Levin and Straat using the lunar soil, the LR response was typical of sterile soils, with little difference between unsterilized and sterilized lunar samples.

Dr. Norman Horowitz began to analyze his initial results from the soil sample taken on SOL 8 in the PR experiment after a five-day incubation period under Martian conditions, with carbon monoxide and carbon dioxide labeled with carbon-14 mixed in the test chamber with the Mars atmosphere. As already mentioned, the Viking PR

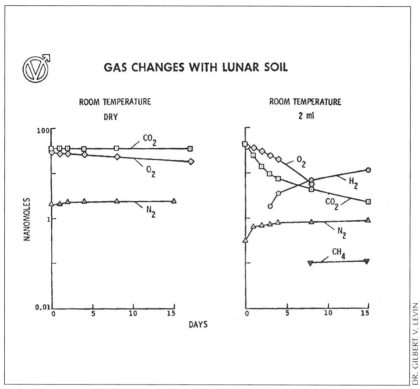

GAS CHANGES WITH LUNAR SOIL

DR. GILBERT V. LEVIN

Oyama's GEx experiment result from a pre-Viking lunar test soil in which he obtains a reaction indicating life! The decline in oxygen and carbon dioxide, and the rise in hydrogen, show changes the GEx considers evidence of life in lunar soil.

experiment did not detect microorganisms directly. Instead it measured the organic material formed by microorganisms in the sample incorporating carbon dioxide and/or carbon monoxide from the atmosphere. For the purpose of the experiment, these gases were tagged with a radioactive isotope, carbon-14. The PR had the option of injecting a small amount of water on the soil sample, but this would be utilized only once with the *Viking Lander 2* at Utopia on Mars. A six-watt high-pressure xenon lamp simulating the Sun's rays was used to "grow" any photosynthetic organisms, and after 120 hours of exposure the unfixed atmosphere that was not incorporated by microorganisms was "flushed" from the test chamber by heating and

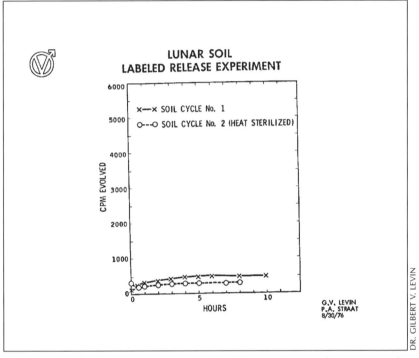

The same NASA lunar test soil run in the LR experiment, showing it to be sterile.

flowing gas. Then the PR temperature was brought to 625 degrees Celsius to burn any organic material that may have formed, thereby releasing radioactive gas. This radioactive gas then is measured by a radiation detector.

A background radiation count of 15 cpm (counts per minute) was considered normal for the PR. In Earth tests, thousands of counts per minute evolved. Horowitz was flabbergasted when his result of 96 cpm came rolling out on his data strip for the Mars test. While the result was low, Horowitz concluded "organic synthesis from atmospheric carbon (carbon monoxide and carbon dioxide) is occurring in the martian soil sample." He continued, "though small by terrestrial standards, it is still significant."

Because the PR was primarily a dry-mode experiment, surface-catalyzed reactions coming from any soil oxidants should not occur,

Dr. Norman Horowitz (left) poses with NASA's Director of Exobiology (in 1976), Dr. Richard S. Young.

so what was being measured by the PR? According to the data coming out of the PR, Horowitz found organic material formed from the simulated Martian atmosphere. No soil oxidants such as carbon suboxide, ozonide, or peroxide in the soil could be causing this reaction! Had Horowitz also detected living cells in the soil of Mars? Dr. Gilbert Levin recalls:

> There were two problems with the PR experiment. One, it could not detect life on Earth unless heavy populations of photosynthetic organisms were tested. Second, the only way that it could be used at all was to filter out the sunlight frequencies below 300 nanometers, because even without organisms, organic material would form! Horowitz published this finding in the *Proceedings of the National Academy of Sciences* [Vol. 68, No. 3, pp. 574–578, March 1971]. He actually proved how organic material must be formed on Mars. He pointed out that over the period of geologic time, significant accumulations could occur. While Horowitz's 96 cpm PR result was significant above background radiation, it was not significant for life. The 300-nm filter in the PR did not really filter out all the energizing rays because subsequently, PR co-experimenter Dr. Jerry S. Hubbard from the Georgia Institute of Technology demonstrated with sterile minerals and rocks that the flight-type PR produced organic material. He simulated the results

that he and Horowitz got on Mars. The organics were condensing on the particles and on the walls of the PR chamber. When they heated (pyrolyzed) those particles, they gave a positive result.

The PR instrument at the *Viking Lander 1* site began to develop a small leak in its radiation counter and thereafter produced unreliable data.

SOL 14

Dr. Klaus Biemann and Dr. John Oro (University of Houston) and ten other scientists representing the GCMS Molecular Analysis Team were waiting for the surface sampler arm of the *Viking Lander* to drop a soil sample into the GCMS funnel. Again the GCMS monitor in the JPL control room did not indicate a sample had been delivered. Beimann was furious. Was the sample in the GCMS chamber or not? Rather than chance heating an empty GCMS oven, Dr. Beimann waited until SOL 22 and purged whatever material might have gotten into the GCMS. The purged material would exit through a small port located on the side of the *Viking Lander*. Beimann and his team hoped to use the Lander cameras to see if the purged soil sample lay on the ground next to the Lander. However, the Lander body obstructed the field of view.

SOL 31

The surface sampler arm dug into the Martian soil about 3.5 centimeters deep. Two hours later a sample was presumably placed into the Viking GCMS experiment. Whether or not the sample was in the chamber, the molecular analysis team decided to run it anyway. When the data strip came in, all it showed were the contaminants that had been in the instrument since its launch from Earth, some carbon dioxide (which was in the Martian atmosphere already), and water. In fact, the GCMS detected large amounts of water when the "sample" was heated. Could it have come from ice crystals deposited or formed in the intake (funnel)? It also detected carbon dioxide. Could this have been dissolved in the ice? These kinds of questions were ignored largely because the GCMS had so accurately recorded the Earth contaminants or the cleaning fluids on Mars.

In spite of the four GCMS organic analysis experiments that would be run on *Viking 1* and *Viking 2*, the GCMS never revealed any other materials than already mentioned. How could the PR experiment detect organic material being formed under Mars conditions if the GCMS—which was supposed to be sensitive to parts per billion—could not? Had the palladium-silver hydrogen separator become poisoned by large amounts of sulfur in Mars' soil? If so, this would have incapacitated the gas chromatograph. However, since the GCMS performed numerous atmospheric analyses and obtained quality data it meant that at least the mass spectrometer portion of the GCMS was working. Was it possible that frost, which is deposited every night on Mars, got deposited on the intake funnel of the GCMS and prevented any soil samples from entering the test ovens? If this was the case, the mass spectrometer portion of the instrument would still be capable of sampling the Martian atmosphere contained inside the instrument.

Even with all of these questions unanswered, one fact remained: the GCMS required the organic material equivalent to that from 10,000,000 microorganisms to get a response. Pointing this out, Biemann said the GCMS would have detected the millions of dead organisms that would have accumulated in the soil if life were present. Thus, it was not a life detector, but a detector of organic matter, living and dead.

SOL 36

Levin and Straat now injected a new soil sample with a drop of their radioactive nutrient and incubated it in the dark for seven days. The reaction was similar to the first, with the radioactive carbon dioxide gas rising steeply at first and within a day slowing down and reaching a level of 10,000 cpm by the seventh day. At that time, the LR team provided a second injection of nutrient into the same soil sample. Everyone in the JPL control room expected to see a second peak of radioactive gas evolve from the sample. After all, if hungry organisms "ate up" all the first nutrient, they certainly would like a "second helping." However, a surprise occurred. No more radioactive gas was evolved from the sample. Instead, 20 percent of that already evolved rapidly disappeared! It must have been reabsorbed by the freshly wetted soil. The scientists who pored over Levin and Straat's new data became doubtful that microorganisms would give up the chance for a "second helping" of nutrient. Upon seeing the data, many flocked over to the Vance Oyama "oxidants in the soil" camp. Did the LR have some kind of malfunction? The response the LR got in Earth soils had shown that as few as 50 bacteria or less could give a response. While waiting out the test period, many wondered and speculated what the second injection would do. Not one expected there would be a decrease in the radioactive gas in that test chamber. Levin tells the story:

> When that second drop of nutrient was put in, we saw about 20 percent of the carbon dioxide in the chamber disappear from the atmosphere into the soil, into something down and out of the radiation detection chamber. Over a period of days and weeks, that gas which had been taken out of the atmosphere gradually was returned to the atmosphere. It was back up to where it had been and then it seemed like this very slight rise continued. That was very different, and one of the legitimate comments made was, "If that was life, why didn't the second shot of nutrient produce a renewed response?" My immediate answer to that was, "I don't know, but if it's life, it doesn't have to be like Earth life." It could be that over that eight-day period away from their environment and closed up in a box at uniform temperatures the organisms died. Then when we put the second shot of nutrient in, all it did was provide an aqueous phase for the carbon dioxide that had been evolved by the organisms to redissolve in, so we had a solution forming that pulled the carbon dioxide out of the atmosphere. Then as time went on, that little damp spot made by the second injection of nutrient equilibrated throughout the soil chamber and got so dispersed that it surrendered the carbon

dioxide that had been dissolved and the carbon dioxide then came back out and equilibrated to where it had been before.

We had run an experiment with lichen, in which we found that lichen produce a response very similar to what we had obtained on Mars when a second drop of nutrient was injected after the eight-day period. We saw a reduction in the amount of radioactive labeled gas in the head-space of the LR. It was not as dramatic as the 20 percent reduction on Mars, but it was a slight reduction, approximately 10 percent.

We had obtained many soil samples on Earth that reacted very similarly to the soil on Mars. The most similar response we have seen is when a California soil was put into a flight-type LR instrument in the Mars simulation chamber (at the NASA Ames Research Center) for three days under Viking-Mars LR experiment conditions. The response was very similar to the response we obtained from Mars. The control response was also very similar. However, there is one significant difference between the responses that we got on Mars and the responses we typically get on Earth and which we also get in the Mars simulation chamber: that is, after the seven or eight days—the initial growth period—we inject a second dose of radioactive nutrients. On Earth and in the Mars simulation chamber, we nearly always immediately see a second pulse of radioactive gas evolving from the culture because in that time those organisms used up all the nutrient and became starved. Then we add the new nutrient, and they gulp it down and start metabolizing right away. On Mars, that did not happen. Instead radioactive gas already evolved immediately diminished, slowly reappearing over a month or so. That part of our experiment somewhat dampened the elation from the initial injection and the strong response we got.

SOL 230

The most enigmatic and significant result from the *Viking 1* biology testing came from a sample that had been stored inside the LR hopper since SOL 90. During this time, the sample was stored completely in the dark but exposed to the Martian atmosphere, with temperatures that varied between 10 and 26 degrees Celsius. The LR Team could not obtain a fresh sample for this test because the Viking Surface Sampler Team was concerned that the sample arm might be damaged by the colder late-autumn Martian temperature, now reading –58 degrees Celsius outside the Lander.

A first injection of LR nutrient was placed on the stored sample, which showed a short initial rise in radioactive gas coming off but by far too small to be significant for life. Then three hours later, a second liquid nutrient injection was applied to the sample and radioactivity measured at 16-minute intervals but no increase occurred. After 200 hours of monitoring it was concluded by the LR Team that storage in the hopper for 136 SOLs and in the dark had severely affected whatever was inside the test cell, be it biology or chemistry. However, it did not seem likely that chemical oxidants could be responsible since the Mars agent was so heat-sensitive. But biology, possibly photosynthetic organisms, might perish in the dark sequestered from sunlight and their environment.

Viking Lander 2

Dr. John Oro, co-experimenter on the Viking PR experiment, was quite convinced that the biology results were caused by oxidants on Mars. As each GCMS organic analysis showed no organics "at the parts per billion level," it seemed to him the only explanation. Even though Mars only receives 43 percent of the Sun's energy that Earth does (because of its distance from our star), it still gets more ultraviolet radiation than Earth (10–20 percent more) because of Mars' thin atmosphere. Oro proposed that hydrogen peroxide was forming in the atmosphere of Mars by the process of UV rays impacting and splitting water molecules into hydroxyl radicals which combine to form hydrogen peroxide.

It was Dr. Oro who embraced Vance Oyama's theories of oxidation on Mars and carried them on into the laboratory. Oro got permission from Drs. Harold P. Klein and Donald DeVincenzi to use the Mars Ultraviolet Simulation Facility at the NASA Ames Research

Center to expose some organic compounds to UV. Dr. DeVincenzi was head of the exobiology department there and was interested in determining whether these reactions on Mars were taking place. The experiments confirmed that the simulated Mars UV light degraded organic material, as UV was known to do. The experiments also demonstrated what Dr. Norman Horowitz already knew—that UV light shining on water, carbon dioxide, and carbon monoxide formed organic material. However, hydrogen peroxide was not subjected to the UV. Had it been, it would have been the first destroyed.

A simple organic compound, naphthalene, was found to exist in some lunar samples brought back by the Apollo astronauts when analyzed by a full-laboratory-size GCMS. The organic material on the Moon was brought in by meteorites eons ago and exposed to the vacuum of space as well as the intense solar UV and cosmic radiations. It did not make any sense that Mars would not have this same meteoritic organic material, but according to the Viking GCMS experiment, it did not.

The idea behind the hydrogen peroxide theories was that over vast periods of geologic time, Mars would have literally had its organic material "bleached" from the surface. A logical question at this point might be: Earth has water, carbon dioxide, and carbon monoxide in its atmosphere and has solar UV coming into the atmosphere as well, so why hasn't organic material on the Earth been bleached away by hydrogen peroxide, just as is thought by some to occur on Mars? The science of atmospheric photochemistry was still in its infancy at that time and no one bothered yet to look at hydrogen peroxide concentrations on the Earth, which by comparison are on the level of thousands of times greater than could be formed in the atmosphere of Mars. The 3 percent hydrogen peroxide that is purchased in local pharmacies for a disinfectant is thousands of times more concentrated than the peroxide thought to be in the Mars soil. In fact, because the Earth's atmosphere has a substantial ozone layer to absorb UV light from the Sun, it actually shields hydrogen peroxide from being destroyed. Thus the Earth has a much larger inventory of peroxide in clouds, soil, and ice than Mars could ever conceivably have.

Even though these heavy concentrations (compared with Mars)

exist in Earth soils, they have never caused Levin's LR life-detection experiment to give a false positive on Earth soils! Why wasn't anyone in the NASA scientific community looking at that? Oxidants by themselves are not a deterrent to life. So if the Earth has thousands of times more oxidants than does Mars, why hasn't all the organic material on our planet been bleached away? The answer to this question is because of life! If life on Earth were to cease suddenly, the oxidants on our planet would eventually "eat up" all the organic material. However, because life is reproducing organic material at such an overwhelming rate it wins the race. The atmosphere of our world is ten thousand times more oxidizing than Mars. A testimony to this disequilibrium on Earth is fire. Fires are due to all the organics (fuel) and Earth's major oxidant, oxygen. If left unchecked, fires could burn up all the organic material on Earth, providing an analogy for how peroxides and oxidants are thought to react on Mars, if they are proven to be there. So far, no hard evidence exists for them.

THE *Viking Lander 2* AT the Utopia Planitia site would, like *Viking 1* on Chryse, begin the long series of photographs, GCMS experiments, biology experiments, and numerous other scientific work. There was a difference here, though. Utopia temperatures were warmer than Chryse during the night because it was much further north. Why? The summer Sun stayed in the sky longer, since it was northern summer, just like Antarctica on Earth has extended periods of sunshine with the Earth's southern axis tipped toward the Sun. On Utopia, unlike Antarctica, there are no vast snow or ice fields to reflect the Sun's radiation back into space, so you have more or less a build-up of solar energy. Also, keep in mind that since Mars is about twice the distance from the Sun as Earth, the seasons are twice as long. Northern spring on Mars represents about 29 percent of its total year and northern summer about 27 percent.

The average surface temperatures at the *Viking 1* Chryse site during the biology sampling activities, as measured by the sensors on the collector head of the sampler arm, ranged from a high of –10 degrees Celsius to a low of –83 degrees Celsius, whereas the *Viking 2* site at Utopia was –32 degrees Celsius for a high and -90 degrees

Celsius for a low. While these temperatures reflected the thermal behavior of the surface, I have mentioned previously (Chapter Two) that the Viking Orbiter Infrared Thermal Mapper (IRTM) experiment made measurements of the surface from orbit and found that in areas with a large concentration of rock material, the temperature of rocks and dark soils could reach +37 degrees Celsius (98 degrees Fahrenheit). If you recall from Chapter Two, Dr. Henry J. Moore of the Physical Properties Team found that temperatures measured just under the soil by the *Viking* footpad sensors indicated that the temperature of the sub-soil rose significantly just before dawn. In fact, these temperatures were compatible with the presence of liquid water. [Author's note: This will be covered in Chapter Nine, written by Dr. Gilbert V. Levin.] Utopia was also found to have more water in the soil (almost twice as much) because of its close proximity to the northern polar ice cap. These temperature determinations led Viking scientists to one of the greatest discoveries in the program—confirming that the northern polar ice cap is mostly water ice instead of frozen carbon dioxide, as had been previously thought. The atmospheric measurements around the northern polar ice cap were around −68 degrees Celsius, far above the condensation point of carbon dioxide, which is −150 degrees Celsius.

The images from the *Lander 2* cameras showed a striking rock- and boulder-strewn landscape, much more rugged than the Chryse site. In later efforts to turn some of the smaller rocks over, they were observed to have a water-line ledge of surface material stuck to them, as if they had been slightly submerged in a shallow pond over time (Moore, H. J., 1987). To make this situation even more interesting, a number of troughs crisscrossed the *Viking 2* landing site. Most of the Viking scientists stubbornly refused to believe these troughs could be caused by ancient liquid water and instead attributed them to complex geological activity.

The Utopia landing site had another interesting feature: the numerous rocks with pits in them, more than 90 percent of the rocks within view of the *Viking Lander* cameras. These pits ranged in size from a few millimeters to several centimeters across. First impressions by Viking geologists were that they resembled breccia on Earth—

volcanic rocks with gaseous vesicles. After careful examination, though, the geologists felt that the vesicular character of the rocks was disquieting by terrestrial analogy because of the extreme number of vacuoles in the rocks. These vesicles or "pits" have never been fully resolved. On Earth there are similar rocks in the deserts of Egypt, but the "pits" are few and there is a lot of sand there. On Mars, one of the most peculiar discoveries by the Viking Physical Properties Team was that Mars did not have any sand-sized particles at either landing site. The particle sizes found by both *Viking 1* and *2* were on the order of microns, indicating a very fine silt. This type of material never could cause the kind of erosion that was apparent on the Martian rocks, even with winds of up to 110 mph (which has been calculated based on movements of dust storms across the surface of Mars).

An alternative explanation for active volcanism causing pits in rocks was the process of saltation. Essentially, saltation is the movement of sand-size particles by strong winds, pelting rocks and stones for thousands of years, wearing them down. Dust alone could not cause such pits, and if sand were not present, then how were the rocks being pitted? Mars has ice-fog crystals. Perhaps the constant bombardment of ice on the rocks over eons had created them. But was there a biological explanation as well? Could rocks be dissolved slowly by acids resulting from the microbial metabolism of Martian organisms? The small pits or vesicles created by this biochemical dissolution of rock then can become a comfortable home. I will cover this concept more thoroughly in Chapter Eight—which looks at a possible Martian microbial ecosystem.

More disturbing than the numerous pitted rocks was the lack of pebble-sized material. Both Viking landing sites showed what looked like pebbles embedded in and covering the soil. Yet when the surface sampler arm tried to pick them up, they disintegrated into fine powder. How could extensive wind erosion be taking place on Mars if the rocks did not crack or spall and leave behind pebbles and smaller fines as a result? What were they being held together by?

Levin and Straat Test Utopia

The LR made four tests at Utopia, which would all have a significant bearing on whether the response was biological or chemical or perhaps even both. *Viking 2* had landed more than a month later than *Viking 1*, on September 3, 1976. However, the *Viking 2 Lander* first day was also called SOL 1, or Martian day one.

On SOL 8 the Viking Biology Team got their first samples. The LR experiment got a radioactive count rate of 3,146 cpm within the first two hours of the injection of 0.115 cc of radioactive nutrient on the Martian soil sample in the test chamber. This then rose to 11,192 cpm within 200 hours from the first injection of nutrient. The LR response was 30 percent stronger at Utopia than at Chryse.

The next objective for Levin and Straat was to heat a portion of the same soil sample to 50 degrees Celsius (122 degrees Fahrenheit),

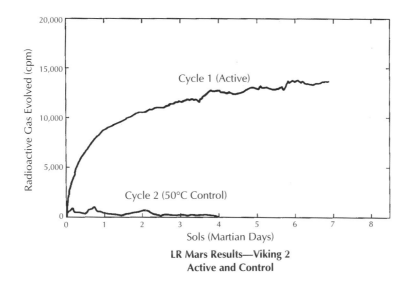

The LR got a 30 percent stronger response at the Viking 2 landing site, as shown in this graph. However, they decided to raise the temperature to only 50 degrees Celsius to see if they might still get a partial reaction. Though reduced by over 80 percent, the reaction was significant. This meant some of the Martian microbes survived at this temperature. Of course, NASA said it could be explained in terms of "chemistry," not life.

instead of the prescribed control of 160 degrees Celsius, to see if a more moderate temperature had an effect on the "agents" in the soil. The lower the temperature that reduced the response, the less likely it was caused by chemicals. This sample was taken on SOL 28 and heated for three hours prior to a nutrient injection. Engineering data returned from the cell showed the temperature reached was 51 degrees Celsius. The soil quickly responded with a short burst of activity which was even more quickly re-absorbed. This repeated several times before leveling off near zero, indicating that whatever was in the soil was sensitive to even the 51 degrees Celsius response and "died."

Dr. Harold Klein simply did not believe the results were valid. If they were, it would put heavy constraints on the chemical theory because none of the suggested ozonides, peroxides, or carbon sub-oxide could be destroyed at this temperature. So what caused the flat-line? Klein and others suggested a malfunction of the LR instrument. Dr. Gilbert Levin recalls the events of that test:

> TRW Space Technologies Laboratories of Redondo Beach, California, who built the flight version of all the biology instruments, ran extensive tests and reported that no instrument failure had occurred. The data from SOL 28 were real! The Biology Team did not even want to discuss it with me. The idea was we were aiming for 50 degrees Celsius as a control—this was all *ad hoc* after people started questioning the results. I got the idea that maybe we can do a better job of discriminating between biology and chemistry than the 160 degrees Celsius control done at *Viking Lander 1*. The Biology Team had all agreed before the Mission that the 160 degrees Celsius heating step was a good separation point between biology and chemistry. But now they thought that maybe the chemistry was being destroyed at 160 degrees Celsius. However, they all selected the 50 degrees Celsius heating, and all agreed that if the response in the soil was significantly damaged, it would certainly indicate biology over chemistry.
>
> We then asked the engineers of the Surface Sampler Team if they could set up a 50 degrees Celsius heat sterilization run for the LR, and they said they could. When we analyzed the data, we found that we actually had a 51 degrees Celsius temperature. Investigating the idea that the positive responses of the LR may

have been caused by UV light activating the soil, we asked if a rock could be moved so we could take a sample under it—where no UV had shone. This revealing LR test came on SOL 51, when the Biology Team would all get a sample from under a Mars rock using the surface sampler arm to dig underneath. The idea here was extremely significant. Why? The rock was most likely there for millions of years, protecting the soil underneath from UV radiation and also from the accumulation of any putative hydrogen peroxide or carbon suboxide forming on the surface of Martian soil. The engineers sent the sample arm out just before dawn and moved a rock. Then quickly, before the dawn intensified, a sample was taken from that spot—which geologists said had been covered by the rock for hundreds of thousands of years. The LR experiment on that sample was positive, less so than the first sample at *Lander* site 2, but equal to that at *Lander* site 1.

Levin and Straat decided to go for the 50 degrees Celsius control again. This time the cell temperature reached only 46 degrees Celsius—and made for a vastly different result. A positive response was obtained, but it was only half as large as the positive produced from under the rock. What did it mean? Levin says, "What we have is the possibility that we are on the threshold of killing the organisms and have not quite done it at 50 and 46 degrees Celsius; they are still living. They are reduced in number or impaired but still living." There is an excellent parallel for that on Earth where *E. coli* is separated from fecal *E. coli* in a mixed culture. Both groups of organisms survive at 37 degrees Celsius. Raise this temperature to 44 degrees Celsius and only the fecal *E. coli* survive. The other coliforms die.

So what happened when the GEx and PR got a sample from under this rock? Since the GCMS also got a sample from under the rock, let's go over what it found first: again nothing. Maybe some water, according to Dr. Klaus Beimann, and the same carbon dioxide data he had recorded at *Viking 1*. He also got readings for his Earth cleaning fluids still in his machine. The GEx data showed about the same amount of oxygen as the other soil tests on *Viking 2*, which were less than half of the *Viking 1* site. The question then was raised, if peroxides were formed on the soil, then how did they get under the rock which had been there for thousands of years? The Viking scientists favoring chemistry said it diffused there from nearby. Dr. Norman Horowitz and his PR experiment never again got the weak positive results he received initially at the *Viking 1* sampling site. He got very weak positives, but still above the radiation threshold of 15 cpm.

The Viking XRFS Inorganic Chemistry Team also analyzed a soil specimen from under this Martian rock (named Notch). Their results? There was less sulfur and much less iron under the rock than the exposed soils. Chlorine and bromine were still present in the sample from under the rock. The fact that there was substantially less iron under the rock was a paradox. Could there be a biological explanation for such a finding?

Dr. Carl Sagan wrote an interesting scientific paper in 1973 entitled "Ultraviolet Selection Pressure on the Earliest Organisms" (*Journal of Theoretical Biology*, Vol. 39, pages 195–200). In this paper he says that water or ice containing sufficient impurities could act as a UV shield. Recall from Chapter Two Dr. Ron Hoham's work on snow-algae such as *Chlamydomonas nivalis*—these organisms store dust and metals within their cell structure for use as nutrients and as protection from solar UV, in addition to the unique pigments they contain which also protect them from harmful solar UV rays. Other researchers found that the smallest particle of iron could protect a microorganism from UV. Martian soil contained 16–20 percent iron by weight in the surface fines. Could it be that some of the Martian organisms without highly developed pigments to protect them from UV attached themselves to a tiny fleck of intracellularly produced iron or magnetite as protection while being blown about by Martian winds? There are thousands of different species of bacteria, algae, and fungi being carried around in Earth's atmosphere. Some hypotheses of life on Earth even propose that life may have begun in the atmosphere of our planet. If this were true, like Mars, Earth had little or no ozone layer billions of years ago and microorganisms being blown around and floating in the early atmosphere would need some form of protection from solar UV rays. Perhaps the microorganism technique of biomineralization to produce extracellular and

170

intracellular iron, magnetite, and sulfur has its origin with organisms evolved to include a "life in the atmosphere." Was this the reason that substantially less iron was found under Martian rocks? Obviously, any living microorganisms under Martian rocks would be protected from UV and so would not need to produce or carry the "fleck" of iron or magnetite with them.

Recent research concerning damage to cells from ultraviolet light has shown that it is much more damaging in the presence of oxygen. Since the Martian atmosphere is 95 percent carbon dioxide with trace amounts of atomic oxygen, UV-oxygen damage would be substantially reduced in any living cells there. Ancient photosynthetic microorganisms using light energy and ferrous iron on Earth could have constructed their own external UV shields, so why not on Mars? Evidence of both ferric and ferrous iron is present in Precambrian sediments and in the Isua sedimentary rocks of Greenland, considered to be the oldest known (3.85 billion years). On Earth, communities of microorganisms formed microbial mats, or layer upon layer of dead and living cells, about 3.5 billion years ago that existed in shallow waters. The uppermost layers of these microbial mats contained substantial amounts of ferric and ferrous iron. So matting itself is a way microorganisms can protect themselves against high UV flux rates when there is no ozone layer protection in a planetary atmosphere.

It was only after the emergence of oxygen in our atmosphere that a global ozone layer built up and started to absorb UV radiation, permitting large land animals and plants to evolve. Prior to this, only microbes inhabited Earth. In fact, microorganisms were on Earth for 80 percent of the entire geologic history before any other life forms evolved, and possibly even longer, depending on which theory of the origin of life on Earth you prefer—Oparin-Haldane or Panspermia.

Another interesting point regarding the snow-algae relates to the fact that they are eukaryotes—oxygenic photosynthetic organisms with a developed nucleus that evolved on Earth only after the prokaryotes, thought to be the most primitive organisms (without a nucleus). The argument is that snow-algae evolution would have had to wait millions of years for an oxygen atmosphere to develop before becoming abundant on Earth. The relevance of this to Mars, of course, is that Mars has only a carbon dioxide atmosphere. So most scientists today would say because of that, Mars cannnot possibly have evolved eukaryotes such as snow-algae because there is no oxygen on Mars. If, however, the theories of hydrogen peroxide formation in the atmosphere of Mars are found to be correct (that it is produced similarly in Earth's atmosphere), then we may have a mechanism for scavenging oxygen and water from the soil of Mars (as discussed previously). So here is a marvelous survival tool that could easily provide aerobic organisms with the oxygen they require without any mass production of oxygen in a planetary atmosphere, and as a bonus have an abundance of water to use as well! If the hydrogen peroxide scenario proves to be correct about Mars, it could have vast implications for how aerobic life may have evolved on Earth billions of years ago before the appearance of an oxygen atmosphere. Scientists are already revising their views on the importance of chemically and biologically produced oxidants such as hydrogen peroxide and superoxides that will alter the way we look at the relationship between biological and chemical weathering processes.

SOL 145

SOL 145 was scheduled as the last and final cycle for the LR experiment on Mars. The surface temperature outside the *Viking* spacecraft was a very cold −84 degrees Celsius! It was late autumn on Mars and time to conclude the biology tests before Martian winter, due to the hazard extreme low temperatures presented to the sampler arm. The LR's test cells already had other samples in them. So Levin and Straat decided to place a new sample on top of one used previously.

Before the test could be performed, the *Viking Lander 2* had its scheduled winter power shutdown. The LR would now have to wait for eighty-four SOLs before it could run a test, until Martian spring. When it was finally time to run the test, Levin and Straat injected a drop of radioactive nutrient on the soil sample that had been stored in the LR hopper for eighty-four SOLs. Immediately some radioactive gas evolved to 350 cpm and went flatline. Compared to the previous LR experiments using fresh samples within a few days, this stored sample again proved, just like that of the *Viking 1* SOL 91, that long-term storage diminished the LR reaction.

What chemicals on Mars could decompose in the dark, Levin and Straat asked? Although they presented their data to the Viking Biology Team showing that heating Martian soil samples to 160 degrees Celsius stopped the LR reaction, that 51 degrees Celsius nearly eliminated most of the reaction, and that 46 degrees Celsius only partially "killed" the reaction, the chemical oxidants were still considered the cause. The fact that long-term dark storage in the LR for months reduced the reactions by 87–90 percent was by itself cause to re-examine "evidence" for life, not oxidants, on Mars. Evidence is distinct from proof—evidence is something to be considered when trying to determine whether proof exists. There was clearly evidence for life on Mars discovered by Levin and Straat.

After much discussion of the biology data, Biology Team leader Dr. Harold Klein made the official NASA announcement based on the evidence supplied by the GCMS along with the unproven hypothesis of peroxides on Mars that "no evidence" of life was found on Mars. The line "no evidence was found by the Viking Biology experiments" would become a campaign to those within the NASA scientific community. How could dedicated scientists ignore "evidence"?

Strangely, the oxidant hypothesis of "tricking" the biology experiments would in time become incredibly complex, involving at least three different oxidants whose composition could not be certain. Some scientists say that hydrogen peroxide is the only candidate that could come close to reproducing the Labeled Release experiment results and would need to be present in the soil of Mars at 30 parts per million (Clark, 1979). Even more bizarre is the fact that these same scientists proposing the oxidant hypothesis know that Earth organisms would have no trouble living in the small quantities of oxidants in Mars or Earth soils! As I mentioned before, Levin tested his LR instrument on Earth hundreds of times, and in spite of the fact that Earth is ten thousand times more oxidizing than Mars, he never failed to get a positive reaction for life in his experiment when organisms were present.

Yet, Dr. Levin and Dr. Straat would now become the focus of countless personal attacks by NASA and eventually be labeled "Percival Lowells" for their continued persistence that their data implied life, not chemistry. Levin and Straat, however, would continue scientifically eliminating one by one the chemical possibilities for peroxides on Mars.

It is interesting to note that whenever the media called on NASA for opinions about the possibility of life on Mars, they were only referred to those in the NASA scientific community who upheld the Viking "no evidence" for life on Mars scenario. More often than not, NASA referred media interested in life on Mars either to Dr. Norman Horowitz, who, Dr. Patricia Ann Straat said, was "vehemently against the idea of life on Mars, and in many cases said he would gladly eat Mars soil to prove life didn't exist"; or to Dr. Klaus Biemann, whose instrument, the GCMS, did not find evidence for organic material on Mars. These would be the NASA spokesmen for life on Mars during the next twenty years, whenever the issue was brought up.

CHAPTER FIVE

Aftermath: Violation of the Territorial Imperative

Small patches were selected in a variety of locations: on rocks, on the soil surface, and in the trenches dug by the sampler arm ... no features have been observed which are unique attributes of photosynthetic pigments.

—Dr. Elliot C. Levinthal and Dr. Carl Sagan, *Journal of Geophysical Research*, Vol. 82, No. 28, 1977, page 4472

The biological experiments on the Viking Landers *did not detect any positive signs of life or any of the organic compounds that are so abundant on Earth.*

—from the plaque mounted near the *Viking Lander* Exhibit at the Smithsonian National Air & Space Museum

Wagner ITH THE CONCLUSION OF THE *Viking Lander 2* biology experiments at the Utopia landing site on Mars, most of the biology team scientists went home satisfied that their experiments worked as planned, but Dr. Gilbert V. Levin just could not believe the ease with which NASA seemed to be accepting the "no life on Mars" verdict. He was too skilled a scientist and he knew his LR machine like nobody else in the world.

There were at least two major obstacles Levin and Straat thought needed to be addressed before life on Mars could be written off. The first was whether the GCMS was sensitive enough to detect the organic matter from a sparse number of organisms that could be in the Martian soil, or whether the GCMS had worked at all due to palladium-silver separator poisoning. Secondly, any microbiologist worth his or her salt knew that Mars was the perfect planet to lyophilize organisms. The preservation of biological specimens by rapid freezing and rapid dehydration in a high vacuum is a technique routinely used by microbiological institutes such as the Centers for Disease Control and the American Type Culture Collection to store microorganisms for future use. By the late 1960s it was well known that there was almost no limit to the length of time microorganisms could be preserved in this manner. Dessicated microorganisms had even been revived from the tomb of the Egyptian Pharaoh Cheops that were 2,000 to 3,000 years old.

If Dr. William Sinton's data and theory were correct—positing that in the dark areas of Mars lay most of the organic material in the soil, while the light-colored desert regions were lacking—then the Viking Landers would have expected to find a small inventory of organic matter because both *Lander 1* and *2* were in these light-colored desert regions. Yet in spite of this, the Viking biology experiments found evidence of "activity" in the soil there. If the organisms were not alive in the soil, perhaps they were lyophilized, like the organisms that lay dormant in Cheops' tomb, but blowing around on the surface of Mars until contact with moisture could be made. What's more, many of the NASA exobiologists knew that if a planet such as Mars is impacted by a meteor or asteroid and the soil and dust are mixed with the rock debris or even just a dusting of lyophilized microorganisms on their surfaces and thrown out into space, the survival of a few microorganisms is possible even in the vacuum of space. This is a far more threatening problem to the Earth than the impact of a meteor itself, depending on its size. Why? Because if microorganisms are transported in space from planet to planet and survive to land upon another planet's surface, as it has now been calculated they can, then if they are pathogenic (capable of producing disease), they could have terrible consequences for the biosphere, perhaps altering our planet to their needs and conditions! There are many different theories emerging regarding the extinction of the dinosaurs 65 million years ago by an asteroid or meteor. What if that asteroid or meteor had alien pathogenic microorganisms in it, and several smaller chunks of it broke off and made their way into Earth's oceans or forests? The dinosaurs could have just as easily become extinct that way. Perhaps the rise of mammals and humans was the ironic result of mutations from a pathogen from another world altering the course of evolution.

It is worth mentioning at this point that in 1976 the National Academy of Sciences' internal Space Science Board put together a group of twenty-seven scientists calling themselves the Mars Science Working Group, or MSWG. The purpose of this group was to study the immediate feasibility of bringing back a soil sample from Mars. When this research group began its deliberations in 1977, it was thought that the next Mars mission would be ready by 1984, so their goal became bringing a Martian soil sample back to Earth. One of their scientists, Dr. Donald L. DeVincenzi (who would eventually become head of NASA's exobiology program), published a detailed NASA report recommending the construction of an orbiting quarantine facility to analyze Martian soil samples. The report, entitled the Antaeus Report (NASA SP-434, 1981), concluded that the safest option would be to have a quarantine laboratory in a space station orbiting the Moon. This way, if Martian organisms were pathogenic and a hazard to Earth's biosphere they could be studied without impact to the Earth.

In 1975, long before the Antaeus Report, Gil Levin, Pat Straat, and Dr. Rudolph Schrot published a massive report for NASA, "Technology for Return of Planetary Samples" (NASW-2280), which served as the basis for the Antaeus Report. It also recommended building a biological laboratory in orbit around the Moon to ensure the protection of Earth's biosphere. However, since 1992, this line of thinking has changed and the National Academy of Sciences' Space Studies Board and the MSWG are now considering the return of a Mars sample directly to Earth by robotic spacecraft, entering the atmosphere and parachuting to a yet unknown destination.

Though *Viking* would continue its meteorology-gathering functions and imaging capabilities until 1982, the biology experiments and organic analysis were over by 1977. There is, however, an unspoken code used by NASA science teams when conducting planetary

Viking Lander 1 *and* 2 *continued to image the surface of Mars until as late as November of 1982, providing tens of thousands of surface images.*

missions such as Viking: you may not violate the territorial imperative. While this statement sounds like something out of "Star Trek,"
it is the code by which most all NASA scientists abide. A geologist
stays with his field of geology, an inorganic chemist stays with inorganic chemistry, and so forth, with one exception: Dr. Gilbert V.
Levin. In the tradition of Copernicus, Bruno, and Galileo, Levin
refused to submit to the overwhelming viewpoint of the status quo.
He felt bound by the scientific method, which by any objective measure supported the LR's evidence for life. Being the discoverer of life
on another planet was heady stuff, but the larger meaning was that
we are not alone in the universe, and that was mind-boggling! It was
a discovery that could require all the history books to be re-written,
revolutionizing how humans view the origin of life in the universe.
It bothered Levin that such an important discovery could be so easily cast aside and ignored by scientific symposiums. One might think
that NASA would have convened a symposium just to review and
evaluate the LR data, right? Wrong! Both Levin and Straat tried to
convince NASA to hold such a meeting with all the other scientists,
but it would never happen. For whatever reason, NASA closed its
book on the Mars-life issue. Seeking other possible evidence, Levin
made numerous requests to the JPL Imaging Team leader to analyze
the *Viking Lander* images for a possible biological correlation with
the LR experiment. His requests initially were denied. Gil Levin was
now ready to violate the territorial imperative.

Colors

Frustrated that his LR data had not been accepted by his Biology
Teammates and NASA as strong evidence for life, Levin tried to think
of other ways to corroborate his findings. Finally after months of per-
sistence Levin got permission from JPL to use the Viking Imaging
System facilities, where all Viking images are reposed. Dr. William
Benton, Steve Wall (who had been a co-investigator on the Viking
Imaging Team), Donald Lynn, and their associates readily gave Levin
access and technical help. However, he was told that E. Levinthal,

K. Jones, P. Fox, and Dr. Carl Sagan had already looked for evidence of life in the pictures and had reported no camels or elephants walking across the field of view. Levin was interested in smaller objects so this did not deter him.

In 1977, Levin began a painstaking search of some ten thousand Lander images. He requested that the images be fed to his viewing monitor in radcam. It was at once evident to him that the true color of the Martian landscape was as depicted on the first panorama, which NASA had quickly withdrawn in order to substitute its arbitrary orange-red world. To prove this, Levin and his JPL Viking Imaging System associates used the digital spectrum analysis capability of the system. He was told that this was the first time the images had been examined at this fundamental level. The Viking Imaging Team had viewed the images and selected some for immediate public release and later in scientific journals, but no one had yet performed a detailed analysis of the three color and three near infrared channels of data. The head of the JPL Viking Imaging System told Levin he was glad someone was finally making use of the expensive capabilities that had been built into the system.

While he was at JPL, Levin spoke with Dr. Raymond Arvidson, who had been working with the Lander Imaging Flight Team and who would later, on the Viking Extended Mission, became the Viking Science Team Leader. Dr. Arvidson and a graduate student of his from the Department of Earth and Planetary Sciences at Washington University in St. Louis, Missouri, were at JPL doing studies of the geology of the Viking landing sites. Arvidson's grad student, Edwin Strickland III, was brilliant when it came to computer enhancement techniques and would eventually play a significant role regarding the Mars color issue, but it would be Levin who would start them on their way. Dr. Levin recalls:

> While I was at JPL, Ray Arvidson and Ed Strickland were there doing some analysis of the geology on Mars and looking through the images taken by the landers. I then decided to call their attention to what I had found and said, look, I have found images of rocks with green spots on them and the Martian landscape is not uniformly red. I'm not saying that green means life, but I am saying

that Mars is not monotonously orange-red, as NASA has published. Furthermore, there are some interesting patterns on some of these rocks. Arvidson and Strickland both looked over my findings and said, "hey, that's very interesting," and left it at that.

Arvidson knew the colors of Mars were very different from the NASA-released media images. He along with six other prominent scientists released a paper for the *Journal of Geophysical Research* on September 30, 1977, which said:

> The predominantly yellowish brown color of the two pictures shown in figure 1 departs appreciably from the reddish brown color of pictures that were released for many months after the landings on Mars. The incorrect reddish renditions resulted from initial attempts to balance pictures by reproducing the color of the reference test charts and of the grey paint and orange cables of the landers....

This supported what Levin and Straat had seen in the Viking control room. When the first color image came in, it might have been the correct one. Levin quickly confirmed that there were a variety of colors present in the soil and on the rocks, ranging from brown to ocher to yellow to yellow-green and even green. His interest focused on three rocks with some greenish coloration, located close to *Viking Lander 1.* The largest—roughly shaped like a triangle—he called "Delta Rock," the second was called "Flat Rock," and the third, for a reason that made it of great interest to him, "Patch Rock." This last bore a peculiar, well-defined, greenish patch on its top and side. Whimsically calling the group "the Delta Rock Trio," Levin began examining the three rocks in detail. *(See Color Plate 12.)* Using the spectral analysis of the imaging system, he determined that the greenish coloration was real. Looking intensely at the image, Levin wished the pixel size (the squares of which the picture is made) were smaller to permit better definition. At the distance of the Delta Rock Trio, the pixel size was about one-half centimeter on a side, too large to see details of a microbial colony or other life-revealing level of organization.

During the Viking Mission, and until 1982, the Lander cameras kept taking pictures. Levin was inspired to compare the Delta Rock Trio images taken at different times but at the same sun angles—technically called "repros" (work which Levin is still involved with today). He got the JPL imaging technicians to split the screen of his color monitor so he could compare images side by side. Levin recalls what he saw:

> I will always remember one night about 9 P.M. when I pulled up images of the Delta Rock Trio taken on SOL 1 and SOL 302 and examined them. The greenish patch on Patch Rock had changed! I felt the same heady rush as when the LR data first came in. Too excited to continue work, I left JPL and drove up Angel's Crest Highway well into the hills, parked off the road, and looked up at the stars until I calmed down.

Levin and Straat immediately went to work on their own scientific paper and published it in 1978 in the *Journal of Theoretical Biology*, Volume 75, pages 381–390, entitled "Color Feature Changes at the Mars Viking Lander Site." In this exhaustive study of the *Viking Lander 1* site images, Levin and Straat along with Dr. William D. Benton of JPL concluded that the color of patches on the rocks was green and appeared to change with time. *(See Color Plate 13.)* All possibilities were considered as an explanation, including exotic soil chemistry and minerals, but it definitely left open a biological interpretation as well. Levin and Straat had submitted this paper to both *Science* and *Nature*, the two prestigious scientific journals recognized the world over, but were flatly rejected.

Levin on subsequent returns to JPL continued to review the Viking Lander images and got Dr. Carl Sagan involved in 1981 since he was a member of the Viking Imaging Team. Levin and Sagan spent a few days together doing pixel differencing on images of "Levin's rocks." Dr. Levin recalls:

> Soon afterwards, Carl [Sagan] called me to say he had analyzed the pixel differences, found them to be real, and enthusiastically congratulated me for documenting changes that implied life on Mars. Subsequently, he took it back, commenting that he had now calculated the differences to be significant to one sigma only.

Scientifically speaking, what he meant was the [one sigma only] probability was about 68 percent of it being real. I thought that this stance was odd, because Carl had earlier placed a 10 percent probability of finding life on Mars, saying even this was well worth the effort. Sixty-eight percent is a hell of a lot bigger than 10 percent.

Dr. Levin then sent Dr. Henry J. Moore of the Viking Physical Properties Team a copy of his paper on the colors of Mars and Moore wrote back: "Thank you very much for the reprint of your paper. I think you presented the data and possible interpretations fairly."

The paper fared no better at the hands of the scientific community than had the original LR data, perhaps even worse since many believed Levin was relying on the green color as evidence for life. Levin denied this, pointing out that lichen, for example, came in a variety of colors. His emphasis was on the pattern changes. Many then contended that the "changes" were merely due to shadows created by the different sun angles at which the images were taken. Wondering about the sun angle issue, Levin went back to JPL and pulled up "repros" (images taken at the same sun angle on different Sols) of the Delta Rock Trio. Comparing them would eliminate any possibility that a change in sun angle had made apparent differences. The differences remained, but so did the objectors. Just as in the case of the LR data, the Delta Rock Trio data were largely scorned by the scientific peerage.

A flurry of revealing scientific abstracts (1979 to 1981) would later be presented by Dr. Ray Arvidson's graduate student, Edwin Strickland III, that completely vindicated Levin and Straat's pioneering work and their violation of the territorial imperative. Strickland's first abstract was presented in 1979 at the Second International Colloquium on Mars (NASA Conference Publication 2072). Entitled "Color Enhanced Viking Lander Images of Mars," the abstract reads thus:

> At the *Viking Lander 1*, five distinct soil colors can be discerned. [*See Color Plate 14.*] A bright, relatively red soil forms a thin, discontinuous layer that covers most other soils. This unit is a few centimeters thick near the lander, and may be only microns thick.

183

A darker blue soil is exposed (a) where the bright red soil was eroded by lander engine exhaust, (b) in the Sandy Flats trench (biology sampling site), (c) in irregular patches between rocks... The Rocky Flats sample site, and other patches between rocks and on drifts, show a distinctive green-blue color....

Strickland's next scientific abstract was published in *Proceedings of Lunar and Planetary Science 10*, Volume 3, pp. 1192–1194 (1979), entitled "Soil Stratigraphy and Rock Coatings Observed in Color Enhanced *Viking Lander* Images." It was electrifying and seemed to go unnoticed by the NASA scientific community, except for Levin. Strickland reports:

Subtle color variations of the martian surface materials were enhanced in eight *Viking Lander* color images processed at the U.S. Geological Survey's Flagstaff, Arizona, Image Processing Facility ... color artifacts due to Lander cameras are minor and do not hamper image analysis. [*See Color Plate 15.*] Two tests demonstrate that colors in the enhanced images represent real features of the martian scene. Overlapping images taken with separate cameras reveal the same relative colors. Two images of the same scene taken at different times of day show that colors of sunlit materials do not depend strongly on illumination ... five distinct soil colors have been recognized at the *Viking 1* site ... a thin layer of intermediate brightness "green-blue" soil is present as smooth patches that fill shallow dips and hollows in the duricrust ... Rocks at the *Viking 1*, most probably bedrock and many other loose rocks near the lander, have very light surfaces, with a green ratio greater than any other surface material. They appear bright green in enhanced images... At the *Viking 2* site, only one color population of rocks appears to be present. Green colored surfaces are abundant on rocks of all textures. Many rocks appear to have a thin (millimeters) coating of red soil that is identical to the color of the red drifts and duricrust. The brightest rock surfaces are yellow areas that appear at the boundaries of red and green rock surfaces, and on projections where the red coating appears to be thinner. A speculation is that the yellow material may form on green surfaces that are covered by the red soil.

Edwin Strickland's amazing paper concludes:

Significant color variations occur for soil and rock surfaces on Mars, and these can be used to infer soil stratigraphies and the presence of rock coatings at both sites. The great similarities in relative colors at both sites imply that planetwide and not locally important processes are involved... The observations are consistent with the presence of igneous rocks that develop green weathering coatings, and that may have thin coatings of adhering soil... If the color of rock surfaces does not reflect mineralogical variation within the underlying rock, it can be concluded that they are a result of globally active environmental processes.

As astounding as his work on the colors of Mars might have been, Strickland himself did not feel the colors could be attributed to biology. When Levin read Strickland's papers he immediately called him. Ed Strickland comments on his conversation with Dr. Levin:

I looked through Gil Levin's material and told him I looked at the spectra of the rock and soil materials in the infrared with the *Viking Lander* cameras and they showed nothing unusual. The Lander camera had red, green, and blue filter diodes along with three near infrared channels that were called IR-1, IR-2, and IR-3. These were selected within the sensitivity limits of the camera. We used them to try and pick out different oxidation states among iron compounds because in the near infrared, the typical absorption bands for iron-bearing minerals like olivine and peroxine are different from some of the iron oxides. Chlorophyll, if it were present, ranges from a very low reflectivity in the visible wavelengths and was around 3–4 percent reflective in the red and blue diodes, and a whopping 15 percent in the green, but it's not very bright. The chlorophyll in vegetation goes up to 90 percent reflectence very suddenly in the near infrared. If there had been any chlorophyll on Mars it would have shown up spectacularly in the infrared data.

There are other pigments that can be produced, however, that may not show up in infrared data. The alga, Cladophora, had been shown by Sinton to avoid detection by infrared means. Sinton discovered that this species of alga attaches an oxygen atom to one of the carbon atoms, and this shifts the resonance of a hydrogen atom attached to the same carbon to a longer wavelength. Sinton hypothesized that Martian organisms might store large amounts of carbohydrates as food.

It is also known that certain cells containing bacteriochlorophylls tend to scatter light, thereby reducing or masking the maximal absorbance of their pigments. This would be especially true of an organism on Mars that has developed a "hard shell" or carried a fleck of iron in order to further protect it from UV damage.

Strickland remarked to Levin that his conclusions supported those of Dr. Harold Klein, who said that the Viking experiments do not support speculations on the presence of widespread life on Mars. No one, not even Strickland, who had more photographic evidence to support further biological investigations, wanted to be associated with a biological interpretation of the data. Moreover, the politics of science were so delicate that even though Levin had called Strickland's attention to the greenish spots, Strickland never printed an acknowledgment.

The efforts of both Dr. Arvidson and Edwin Strickland only solidified what Dr. Levin had known all along—that something was wrong here. Why was the NASA scientific community blatantly ignoring crucial scientific data that could support the notion of life on Mars? That's what *Viking* had come for. It was now politically incorrect to even mention life on Mars. If you did, you were branded a scientific outcast. When Ray Arvidson found out Levin published his paper on the colors of Mars rocks, he said sarcastically, "Gil Levin is out of his territory. He's a biologist, not an imaging specialist, and he has no business trying to do imaging analysis." (Personal communication, 1995.) Subsequent publications by Strickland and his group failed to reference Levin's pioneering paper on the colors of Mars.

From the personal files of Dr. Gilbert Levin we read his comments on Dr. Arvidson:

> After discovering three articles in the NASA Reports of Planetary Geology Program, 1978–79, NASA Technical Memorandum 80339, by Dr. Ray Arvidson and his group on the subject of soil color and color changes and rock colors on Mars, I called him

today. I told him I had just discovered the articles last week, had thereupon called NASA Headquarters for an update, and had learned that the meeting of the Planetary Geology Group was in session last week. I explained to Ray [Arvidson] that NASA had sent me its 1980 Technical Memorandum, 82385, explaining that the abstracts therein would be the source of topics for the discussions last week. I told Ray that I had reviewed the abstracts and none had shed further light on the rock spots on Mars, particularly with respect to changes with time in the color pattern of the spots. I told Ray I was curious about the three reports last year and wondered whether he could review the subject for me. Ray, obviously uncomfortable, said that the rock study was done by Ed Strickland, a graduate student. I asked Ray if Strickland was his graduate student and he responded "yes." Ray said that after Ed completed the work reported last year he shifted his emphasis to the *Viking Orbiter* pictures. Ray said that Ed had published his work last year in the *Lunar Science Conference Proceedings*.

I then asked Ray about the soil color changes and he said that E. Guiness (Ray's coauthor in the 1978–79 report) just obtained his Ph.D. on the subject study. I reminded Ray that I had called the subject to his attention originally with our [Levin and Straat's] Patch Rock study at JPL, where I discussed the matter with him. In light of this, I asked why his group had not examined the changes we reported for Patch Rock. He said they were not looking for rock spots at all; instead their goal was to study the soil color changes to see what implications they might have for the deposition of frost on the Mars surface. I told him I thought it was very strange that they should have confirmed my findings of colors on the soil and rocks but not sought to determine whether our detection of changes is also correct. Ray then attempted to explain that the determination of such changes and, indeed, colors, was very difficult and that, in fact, the rocks and spots on them were not "green" but were merely "greener" than the other objects in the field of view. They only appeared green in enhanced color pictures, he said. I told him that is precisely what we had reported in our paper but that the fully radiometric images showed the spots to be green to the eye, and that this seemed confirmed in the second article by Strickland since it referred to green rock surface coatings without qualifying the color as enhanced and

187

also since the areas covered by the green coatings coincided in the enhanced color images and in the six-channel (color filters) study reported by Strickland. I reminded Ray of our earlier discussions in which he had said that he would look into color differencing concerning our Patch Rock changes and that I had been surprised that he had not gotten back to me on the subject of rock and soil colors. I again mentioned that I felt we should have been cited as a reference in those reports and in any publications, and that I was requesting such citation in any future publication. Ray again said, "You have a valid point."

At the request of Ray Arvidson, Ed Strickland III called Dr. Levin on January 19, 1981, and Levin's files summarize the conversation:

> We briefly discussed his work on studying the colors of soils and rocks on Mars. He told me he had published a paper in the *Tenth Lunar and Planetary Sciences Colloquium Proceedings*, 1979, page 3055. He said he would send me reprints. We discussed his work and its relationship to the earlier work we had done and called to Arvidson's attention. Strickland readily acknowledged our earlier efforts and said that he had referenced our paper in the reprint cited above, but not in the NASA reports. Strickland said that he used color enhancement procedures different from ours and found the patch on Patch Rock to be "yellowish-gray." He also stated that Big Joe is a mixture of "green" and "blue" units and that the tips of the rocks are mostly "green." He accounts for the coloration by ferrous weathering yielding a "weathering rind." I asked him why he did not include a biological hypothesis in his discussion. He said that "one must select the reasonable, and Ockham's Razor rules against biology so that it would not be proper to mention it."

The Letter to Carl Sagan

Levin, from his office at Biospherics Incorporated, drafted and sent a letter to Dr. Carl Sagan in December 1980. It read:

> Dear Carl:
> Eureka: You may recall that when I first asked you to look into our Patch Rock data, I told you that Ray Arvidson had said he

planned to do some work to investigate our findings, but it would be some time before he could get to it. I never heard from Ray, but two days ago, I was reviewing the NASA Planetary Geology Reports for 1978–1979 and Voila! Ray and his group have indeed been busy: As evidenced by the three articles therein which I am enclosing for you. In sum: 1. There are green spots on rocks at the Viking Landing Sites—their published picture even featuring our Delta Rock Trio; 2. The spots are unlike any other material in the entire field; 3. The spots seem to be "millimeters" thick; 4. The spots seem to be formed upon the rocks; 5. The IR spectrum of the spots differs from the spectra of any other materials in the field; and 6. Changes in color of the soil around some of the rocks with time have been observed. However, the color reported for the soil patches is blue as opposed to our finding it to be greenish-yellow. Also, no attempt, apparently, was made to confirm changes in the spots on the rocks with time. Of great interest, however, is the statement that the brightening of the later pictures (which effect we also observed) was caused by fallout of several microns of dust which can obscure the spectral characteristics of the underlying material. Despite this, however, our patch on Patch Rock intensified in its greenish color over this time period. Two other things are remarkable in these papers: 1. The effect is attributed to some kind of physical or chemical weathering without any mention being made of a possible biological significance; 2. No reference is made to our paper, published one year earlier and acknowledging the discussions with Ray which put his group onto this work... I am sorry I had to violate the territorial imperative to get the imaging data, but after the Imaging Team denied my request to look for the green spots there was no choice but for me to jump fields ... please do not go on saying nasty things about our publication in the *Journal of Theoretical Biology*. All the best.

Very truly yours,

Gilbert V. Levin, Ph.D.

Sagan's Response

Dr. Carl Sagan wrote back to Levin on February 3, 1981:

> Dear Gil:
> Thanks for your recent letter. I'm sorry I had missed the work by Arvidson and his students. Okay, I admit that there's a prima facie case for something which absorbs at long visible wavelengths and changes with time in rocks at the *Viking 1* and *2* landing sites. I will try in the next year to look into whether plausible photometric functions for the Martian surface can explain short-term changes, and whether aeolian deposition or removal can explain long-term changes. If we find anything interesting I'll be sure to let you know. Meanwhile I promise not to make any derisive remarks about the *Journal of Theoretical Biology*. (I've published there myself!)
> With every good wish,
> Carl Sagan

Journal of Molecular Evolution, Volume 14, 1979

With the conclusion of the Viking biological investigation in 1977, all of the participating scientists as well as other NASA scientists went to work writing the various research papers that would be published in many scientific journals around the world. Of particular interest is the large number of papers that appeared in Volume 14 of the *Journal of Molecular Evolution*, published by Springer-Verlag of Germany. Vance I. Oyama and Dr. Bonnie J. Berdahl, both from the Extraterrestrial Research Division of the NASA Ames Research Center, published "A Model of Martian Surface Chemistry," which essentially promoted Oyama's peroxide, superoxide, carbon suboxide polymer theories.

Also included in this volume was Dr. John Oro's paper titled "The Photolytic Degradation and Oxidation of Organic Compounds under Simulated Martian Conditions," which dealt with the process of how Martian organic surface materials are theoretically destroyed by a combination of UV light from the Sun and oxidizing by-products (hydrogen peroxide) of the Martian atmosphere and surface. Also, Dr. Donald M. Hunten of the Lunar and Planetary Sciences Department of the University of Arizona published his paper, "Possible Oxidant Sources in the Atmosphere and Surface of Mars," again covering the theory that oxidants are the only explanation for the Viking biology experiments.

Dr. Lynn Margulis from the Department of Biology at Boston University along with five co-authors published their paper, "The Viking Mission: Implications for Life on Mars," in which they conclude that the Viking biology experiments are best explained by non-biological phenomena.

Another paper in this same volume was by Dr. A. Banin and Dr. J. Rishpon of the Department of Soil and Water Sciences at the Hebrew University in Israel, entitled "Smectite Clays in Mars Soil: Evidence for Their Presence and Role in Viking Biology Experimental Results." The authors try to convince readers that smectite clays which contain iron and hydrogen as absorbed ions were responsible for the life-like reaction of the Viking LR experiment.

Even Dr. Harold P. Klein, head of the Viking Biology Team, would publish a paper, "Simulation of the Viking Biology Experiments: An Overview," concluding that UV-treated materials, iron-containing clays along with iron salts, and strong oxidants were responsible for the Viking biology reactions.

There was one paper that did not contain these pessimistic views regarding the life-detection experiments on Mars, and it stood alone against all the other papers contained in that particular volume of the *Journal of Molecular Evolution*. [Author's note: This journal can be found at many university science libraries.] This solitary treatise was written by Drs. Gilbert V. Levin and Patricia Ann Straat, entitled "Completion of the Viking Labeled Release Experiment on Mars." They conclude that the reaction they got in the LR experiment was consistent with a biological explanation.

The scientific opposition to life on Mars was enormous. No one wanted to consider biology as an explanation for any of the Viking biology experiments. However, for a full three years, Levin and Straat would undertake the most ambitious study of peroxides ever done in a laboratory to try and simulate the reaction they got on Mars. Their conclusions were published in Dr. Carl Sagan's journal *Icarus*, Volume 45, pp. 494–516 (1981), entitled "A Search for a Nonbiological Explanation of the Viking Labeled Release Life-Detection Experiment." Levin and Straat's work showed that hydrogen peroxide is destroyed at a rate 1,000,000 times as fast as it can be formed on Mars because the UV rays of the Sun break it up. On Earth, ozone protects hydrogen peroxide from being destroyed and it can accumulate in soil, clouds, or water, but Mars has little ozone to speak of in this regard and hydrogen peroxide could not accumulate.

In addition, Levin and Straat demonstrated that hydrogen peroxide is unaffected by a temperature of 50 degrees Celsius, which they used as a control in the LR experiment on Mars. If you recall from Chapter Four what happened, 50 degrees Celsius stopped the reaction on Mars. Using a Mars-simulated UV light environment under Mars atmospheric conditions and pressure, the science team of Levin and Straat could not get hydrogen peroxide to accumulate, making it extremely unlikely that this oxidant was causing the LR to become "tricked" into mimicking a life-like reaction. They also found out why Drs. Banin and Rishpon got a reaction similar to the LR results on Mars using smectite clays from Earth—they were unsterilized samples which contained populations of living microorganisms! Banin and Rishpon discovered Earth clays have life, something scientists had already known for more than seventy-five years. It was embarrassing, to say the least!

Dr. Patricia Ann Straat left Biospherics Incorporated in 1980, saying that government contracts were getting harder to come by and times were lean at Biospherics. She applied to the National Institutes for Health in Maryland and got a position as a Grants Associate, eventually to become Deputy Chief of Referral there. However, Levin and Straat would write one more scientific paper together that would be more revealing about life on Mars than any other previously published paper. It appeared in the *Journal of Theoretical Biology*, Volume 91, pp. 41–45 (1981), entitled "Antarctic Soil No. 726 and Implications for the Viking Labeled Release Experiment."

You may recall from Chapter Three how the Cameron Soil

Samples from Antarctica were distributed among the participating scientists on the Viking Biology and GCMS Teams. In going back over their data base, Levin and Straat found a soil sample they tested in the LR test module on Earth simply labeled "Sterile Antarctic #726." The special aspect about this soil was that it had living colonies of bacteria in it, even though classical microbiological tests had found it sterile. Why was this so special? Because the pre-Viking tests of the GCMS also tested this same sample and determined it was not only sterile but contained no traces of organic matter. Not only had Cameron been wrong again about another Antarctic soil sample being sterile, but the Viking GCMS could not find organic material in it even though it contained life!

The consequences of this were shocking. It proved that the sensitivity of the Viking GCMS was insufficient to rule out small populations of living organisms on Mars, even if the GCMS functioned perfectly, as Dr. Klaus Biemann said it did. In defense of his GCMS experiment Dr. Biemann says:

> The GCMS was never designed to detect the amount of organic material that would be present in a few microorganisms that would be present in a 100-cubic-millimeter sample of soil. That is not the point. The question is, how does one explain that there could be living organisms in an environment that does not contain any detectable organic materials, which would have to be there for the microorganisms to live on or would be produced by the decay of billions of generations of microorganisms that died, unless one develops a 100 percent scavenger theory such as the one Carl Sagan proposed during the Viking Mission. If microorganisms on Mars are so well developed that they can eat up all their predecessors' cadavers, then I can agree completely. However, on Earth, one cannot explain the existence of living organisms in the absence of organic debris. The Viking Biology group prior to the mission all agreed that no apparent positive biological experiment can prove the presence of living systems if there is no organic material in the soil. Furthermore, with all respect to Dr. Gil Levin's LR experiment, he has never proven that he does not get any false positives. (personal communication, 1996)

Both Dr. Levin and Dr. Straat feel that the heat sterilization conducted at 160 degrees Celsius and on Mars at even lower temperatures was proof that their instrument was not getting any false positives. Tests of moon dust and other sterile materials had all been negative. "I really don't understand why Dr. Biemann insists we could not prove we were not getting false positives. It just doesn't make sense at all," Levin states.

During the time Levin and Straat were writing their paper on Antarctic Soil #726, they got a visit from a very excited Dr. E. Imre Friedmann and his wife, Dr. R. Ocampo, at the Biospherics Incorporated laboratory. They brought with them a sample of a rock gathered in Antarctica containing endolithic lichens inside. If you remember from Chapter Three, it was Dr. Wolf Vishniac who sent the first rock samples with the endolithic lichens in them to Dr. Friedmann at the Polar Desert Research Center at Florida State University before Vishniac's death in 1973. Since that time, Drs. Friedmann and Ocampo had been investigating the nature of these organisms that could live inside rocks, and felt they might be good candidates for the type of life that could exist on Mars.

The scientists then decided to run an LR test on the endolithic lichens and carefully scraped them out (so as not to contaminate them) from a 10-mm depth in the rock and put them in the LR to see what kind of reaction they would get. The magnitude of the response was comparable to what was observed on Mars. There was also evidence to suggest that these lichens were able to derive their water needs from atmospheric water vapor only. If this were established as a scientific fact, there would be no reason why endolithic-type organisms or many others could not exist on Mars.

Levin Goes to School

Gil Levin used his 35-mm single-lens reflex camera to take numerous transparencies of the images he viewed at JPL. Home again, he assembled a group of these slides and took them to NASA Headquarters, where a group of key scientists-administrators viewed them.

This meeting included Dr. Richard Young, the Viking Program Scientist, and Dr. Gerry Soffen, the Viking Project Scientist, along with Dr. Walter Jakobowski and three others. Dr. Gilbert Levin recalls the events as they unfolded that day:

> They sat in the room for an hour while I showed them slides of the images I had acquired out at JPL. Suddenly Gerry Soffen stood up and said, "Gil, that's ridiculous, what are you looking at? Do you call that green? That's just gray! There's no green there at all!" And they all began laughing at me. Dr. Walter Jakobowski interrupted the group, stood up, and said, "Wait a minute guys, maybe you're pushing too hard." Again Soffen spoke up while still laughing, "No, Gil, this is absolutely ridiculous and you certainly shouldn't say anything about this."
>
> Completely distraught, I left the NASA meeting with slides in hand around noon and drove out to North Chevy Chase Elementary School in Maryland, which my youngest son, Henry, had attended. I got permission from the school principal to show pictures of Mars to the fourth grade. I hooked up my projector and showed the slides, merely saying they were pictures of Mars. I then asked the fourth-graders what they had seen. Many hands went up. In their descriptions, most said they had seen rocks with green spots on them! My sanity was saved—at the expense of NASA's.

That's all he needed to hear. Levin packed his materials, thanked the class and Henry's former teacher, then left.

Much later, when Henry Levin was old enough to start looking at colleges to attend, he asked his father if he would help him. Once again Gil Levin would have a remarkable school experience, only this time it would prove even more surprising than his grade-school visit. Henry Levin wanted his father to take him to Brown University, where the Viking Imaging Team Leader Thomas Mutch happened to teach. Mutch took Gil and Henry Levin to the Biochemistry Department (Henry was interested in chemistry), where he introduced them to the chairman of the department. Mutch told the professor that this was Dr. Gilbert V. Levin, one of the biologists who had a life-detection experiment aboard the *Viking* spacecraft on Mars.

The chemistry professor turned to Gil Levin and said, "Life-detection experiment on Mars? Now wasn't that an expensive waste of time. The Scriptures certainly tell us there can be no life anyplace but on Earth." This was the Chairman of the Biochemistry Department at Brown University (an Ivy League school)! Levin bit his lip and did not comment further.

1986—The First Viking Anniversary

The Viking Science Alumni gathered at NASA Headquarters in Washington, D.C., on July 20, 1986, the ten-year celebration of the Viking Mission to Mars. Dr. Gilbert Levin was among the featured guests to speak that night. Once again Levin brought slides of the LR reaction on Mars, along with some images of rocks on the Martian surface, and concluded in his presentation that "life was more likely than not discovered on Mars by the Labeled Release experiment." On the way back to his seat, Levin was greeted with hostility from Dr. Norman Horowitz, who said Levin was a disgrace to him and to science. Levin countered with, "When are you going to look at the data?"

The National Academy of Sciences took vehement exception to Levin's announcement, and since then, Levin has found it exceedingly difficult to get funded by NASA or to have scientific papers published. Levin comments: "It just seems to me that there is a rather determined effort, in the words of those in the former Soviet Union, to make me a non-person. Sometimes I think maybe I should just go away and let them do that, but on the other hand, why should I surrender and abandon the scientific method?"

[Author's note: It is the National Academy of Sciences that is responsible for the administrative direction for a Mars Sample Return Mission now being planned by NASA to be carried out perhaps as soon as 1999.]

IN 1992 THE UNIVERSITY OF ARIZONA Press released a huge reference book on Mars simply entitled *Mars*. In this book former Viking Biology Team Leader Dr. Harold P. Klein writes about Levin and

Straat's paper on Antarctic Soil #726 and erroneously concludes that the soil sample Levin and Straat used became contaminated in the years preceding the Viking landings. Apparently Dr. Klein never bothered to check when Levin and Straat ran the experiment: it was pre-Viking, at the same time the Viking GCMS performed its test run on the sample. In fact, Levin and Straat had *another* Antarctic soil sample considered sterile by Cameron—Antarctic soil #664— and this was also tested by the GCMS and found to contain no organic material. The LR readily detected small colonies of microorganisms in #664 as well as #726 (presented here in *Mars: The Living Planet* for the first time). And, despite Dr. Biemann's contention, some rare Antarctic samples were found sterile by the LR.

Klein goes on to say in the book *Mars* that no evidence by the Viking Lander cameras was produced that indicated living organisms might be present, excluding entirely Levin's evidence, of course. The truth is, Dr. Levin and Dr. Straat and other members of the scientific

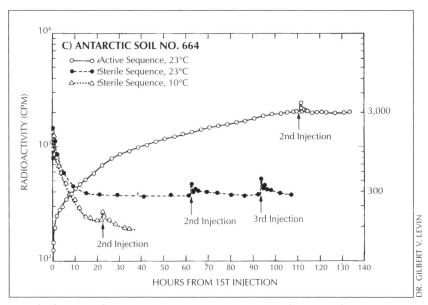

Graph showing the LR getting a positive response from another Cameron-NASA "sterile soil" from Antarctica (soil #664). The Viking GCMS did not find organic material in this soil, either, even though it contained living microbes. (Note that the 2nd injection did not cause renewed gas evolution.)

community for whom that life on Mars is still a good probability were never asked to be represented in this book. The book all but ignores the life-detection experiments aboard *Viking* and instead concentrates on the geology, chemistry, and physics of Mars, even though *Viking* was intended as the search for life. Levin wrote a review of *Mars* which carefully sets the facts straight in the November 1993 issue of the *IEEE Spectrum*.

An Author's Testament

Every now and then, an author writing a book on a special subject actually gets the opportunity to participate in the events unfolding in the book he or she is writing. In the case of *Mars: The Living Planet*, I had two such events that not only helped the research of this book, but also helped me recognize the truth about Dr. Gilbert V. Levin's experiences with NASA.

My interest in Dr. Levin began in 1977 after reading a *National Geographic* magazine that featured the Viking results. In this magazine, the interpretation of life on Mars was very positive. That stuck with me through the years as I started to become a writer myself. I was already writing and being published by 1988, but it was not until 1992 that I was formally introduced to Dr. Gilbert V. Levin. I conducted an in-depth interview with him and published it in a magazine about space called *Final Frontier*. Dr. Levin then had this interview placed among his numerous scientific publications.

The more I talked with Dr. Levin, the more concerned I became that NASA or someone did not want the Mars-life issue resolved. Since I had been using telescopes since the age of eight and had mastered all the basics of astronomy, I thought I might actually have an opportunity to help resolve the Mars-life issue. How? Previous to 1996, the Hubble Space Telescope Science Institute had a program available for amateur astronomers to submit proposals to use the Hubble Space Telescope (this program is now terminated). I submitted a proposal in 1994 and another in 1995, culminating in a research project that took me two and a half years of study on how to use the Goddard High-Resolution Spectrograph on the Hubble Space Telescope.

My proposal was to look for hydrogen peroxide in the atmosphere of Mars using the GHRS ultraviolet spectrometer. Most searches from Earth-based telescopes have been made using infrared spectroscopy because ozone and oxygen interfere with ultraviolet observations. The infrared studies from Earth-based instruments have never revealed the presence of hydrogen peroxide, and infrared research using the *Mariner 9* IRIS experiment did not show it, either. So I thought I would try looking in the ultraviolet spectrum, which, from my studies, seemed to indicate that hydrogen peroxide would show up rather nicely if it were indeed there on Mars. The only way to make useful ultraviolet observations is to get above Earth's atmosphere, beyond the ozone and oxygen.

On my second Hubble proposal I was notified by Dr. Steven Edberg of the NASA Jet Propulsion Laboratory that my proposal was one of five being considered for the final amateur cycle in 1996. I was very excited about the possibility of being able to look for the elusive "not yet discovered" hydrogen peroxide of Mars using the world's most powerful optical telescope. I had also been working on several articles with NASA's Teacher In Space Designee, Barbara Morgan, the elementary school teacher who trained as Christa McAuliffe's backup on the ill-fated *Challenger* mission that killed McAuliffe and the other six shuttle crew members in January of 1986. Barb Morgan always believed NASA should have allowed her to carry out Christa's dream of teaching from space on the space shuttle and was willing to go on a moment's notice. NASA, however, decided otherwise and instead has strung her along as a public relations representative, giving her a space shuttle flight physical every year (the ultimate tease!) at the Johnson Space Center, with no intentions of flying her. I tried to get some national attention on this issue by writing about it in *Quest* magazine. When the Hubble Telescope opportunity arose, I thought I might get Barb Morgan to at least symbolically teach from space using the Hubble Space Telescope with me. I hoped we could link up with the Internet and World Wide Web and teach students across America about Mars exploration. Barb liked the idea but said I needed to get permission from NASA. So I sent in a proposal.

However, it was not to be. About three months later I got a phone call from the Hubble Space Telescope Science Institute that another experiment had been chosen to do a study on galaxies instead. They said only one experiment would be permitted to be conducted, although in previous years, two individuals got to use the Hubble facility. There was some talk for a while that since this was the last amateur cycle, the Space Telescope Science Institute might let all five of us in 1996 do our experiments. An interesting coincidence was that within about twenty minutes of the phone call from the Hubble Space Telescope Science Institute, I got another call from the NASA Headquarters "Teacher In Space Program," saying that my proposal was submitted during the time of the U.S. Federal Government shutdown—a result of the gridlock between House of Representative Republicans and Democrats. I sent the proposal well in advance of this but the point was clear: NASA was not going to allow such a thing.

Previous to that experience I had trust in NASA, but now I wonder. I believe they neither wanted me to look for hydrogen peroxide on Mars nor to have Barb Morgan "teach from space" using the Hubble Space Telescope (sorry, Barb). This experience, however, was not in vain. It taught me about hydrogen peroxide formation in planetary atmospheres and about how scientists conduct research with one another. It also taught me about the politics of NASA. I can now relate on a personal level with Gil Levin's frustrations, except that he has endured more than twenty years of this treatment. Barbara Morgan has endured ten years of it.

The Hubble experience did help lead me to a research scientist who just happened to be making some observations of Mars with the Hubble Space Telescope: Dr. Philip B. James, Professor of Physics and Astronomy at the University of Toledo. Dr. James is assigned to be a participating scientist on the Mars Global Surveyor Mission in 1997 and has led a group of scientists studying Mars using the Hubble Space Telescope. In reading some of the NASA press releases involving Dr. James' work, I became intrigued that the observations he had been making since 1990 indicated that Mars had virtually no dust at all in its atmosphere. So I called Dr. James to discuss with him the colors of the Martian sky. Here is what he said:

When the atmosphere is dusty, the sky color is pinkish: that's because the dust particles absorb strongly at the shorter wavelengths. If the atmosphere is clear, however, it would be more like Earth's atmosphere, blue, except more so because the atmosphere on Mars is very thin. As the dust settles out of the sky of Mars it gets to be a very dark blue or purplish in color. There is not much atmosphere to do Rayleigh Scattering. Normally, the sky is blue because there is quite a bit of Rayleigh Scattering in the Earth's atmosphere, but not so much on Mars, so it probably is pushed to shorter wavelengths and hence the dark blue or purple color.

Then I learned from Dr. James that as dust settles out of the atmosphere of Mars, the sky would undergo a color change ranging from pinkish to yellowish to blue and then finally, after all the dust has settled out, purple. This would also indicate to an observer standing on the surface of Mars that the shadows cast by boulders and rocks stand out in sharp relief, even more so than on Earth. When the sky of Mars is dusty, the shadows become diffuse, just like a cloudy day on Earth.

It was obvious what this meant in the context of this book—that Dr. Gilbert V. Levin, while in the control room of the Jet Propulsion Laboratory with his son Ron in 1976, caught a glimpse of the real colors of Mars at a time when the atmosphere was relatively free of light-scattering particles and before the NASA technicians decided to alter the images. *(See Color Plate 16.)* Furthermore, it must mean that thousands of the *Viking Lander* images on thousands of Martian Sols—the ones with the crisp rock shadows—must all likely have blue skies and a natural brownish soil with other colors mixed in. It is for you, the reader, to ultimately decide why NASA might have done this.

Interesting to note is the October 1996 issue of *Scientific American*, page 20, where we see a newly processed *Viking Lander 2* photograph from JPL archives with brown soil and a blue sky with crisp rock shadows, similar to how that first image must have appeared in the JPL control room in 1976 before NASA Administrator James Fletcher ordered the blue sky negative destroyed and the Martian landscape changed to red.

The truth is, the raw *Viking Lander* images are all on magnetic tape and can be analyzed at any time. All you need to do to get a true color picture of the surface of Mars from this data is to apply the three primary color filters, green, blue, and red, to any of these black and white stored images and you get a very accurate color image of Mars. The Viking Imaging Team almost always included the three infrared diodes on the *Lander* images, known as IR-1, IR-2, and IR-3. This is what gave the Martian landscape its overall orange-red appearance in most of the NASA-released images! You don't need the infrared diodes (filters) for accurate color reproduction—only the three primary color diodes—blue, green, and red.

This same imaging technique is applied in coloring the cosmos. Dr. Harvey T. MacGillivray, an astronomer from the Royal Observatory in Edinburgh, England, has been creating true-color images of galaxies, star clusters, and nebulae. Basically, Dr. MacGillivray takes three images of the same object in black and white—only each black and white photograph is done using one of the three primary color filters—blue, green, and red. The three images are then "stacked" together and printed using digital computer printing equipment. The result is a breathtaking true-color portrait of the universe. It is interesting to note also that all previous NASA spacecraft used the three primary colors to create images of Venus, Mercury, Jupiter, Saturn, Uranus, and Neptune. Why is Mars any different? It's not!

The newly reprocessed images coming out from JPL (of archival *Viking Lander* photographs) on the NASA World Wide Web Site under the heading "Life on Mars" are beautiful—Mars at a time when little dust is suspended in its atmosphere and therefore permitting Rayleigh Scattering, the same principle responsible for the blue skies of Earth. These reprocessed *Viking Lander* images are showing up in numerous sites on the Internet. Why now?

Can it be that with the discovery of the Antarctic meteorites EETA 79001 and ALH 84001 containing evidence for carbonates, organic material, and fossil bacteria that NASA has finally decided to gradually let us in on what they must have known since 1976— that Mars is alive—and they are now planning to take full credit for the discovery of life on Mars without acknowledging the groundwork of Dr. Gilbert V. Levin and Dr. Patrica Ann Straat, who were made a mockery of by the NASA scientific community? (*See Color Plates 17, 18, and 19.*) If this is not the reason, then it simply means that many NASA scientists were wrong and should now openly admit it.

Either way, *Mars Pathfinder* will touch down on the surface of Mars July 4, 1997. On board the *Pathfinder Lander* and *Rover,* will be state-of-the-art color cameras that will show us the changing colors in the skies of Mars as well as the browns, blues, and greens of the soil and rocks (if NASA does not decide to tweak them again using infrared diodes!). This almost sounds like it could be an appropriate place in this book to end our story, but there is much more about Dr. Levin, NASA, and Mars in the 1990s that should be made public.

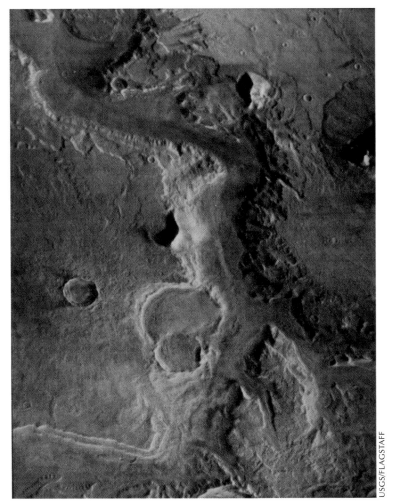

USGS/FLAGSTAFF

Plate 1. *The Reull Valles region of Mars in an enhanced color image to show the wide range of materials on the surface of Mars. Some of the colors are thought to be caused by atmospheric scattering.*

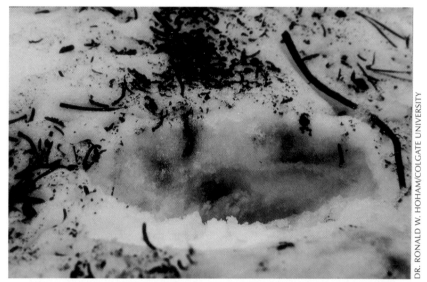

Plate 2. *Green snow-ice algae layered under snow in the state of Washington. Hikers and campers have seen green and red algae coloring large expanses of snow. These come in a variety of other colors including orange, gold, and blue.*

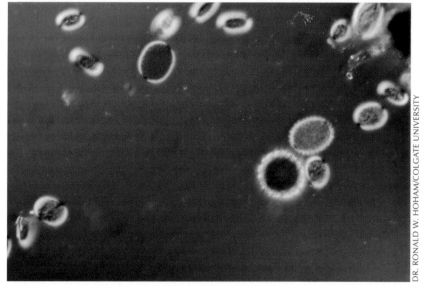

Plate 3. *Spores of red and orange snow-ice algae.*

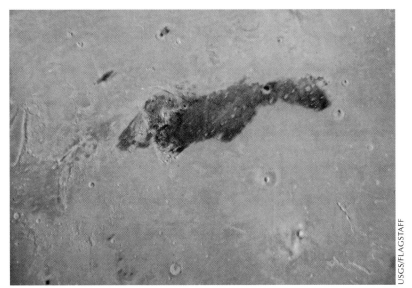

Plate 4. *This isolated dark region in Elysium Planitia is an example of large dark blue-green areas that can form on Mars within months or even days. One such region in 1952 was observed to grow from "nowhere" and was the size of France. No dust storms at the time could account for its sudden appearance.*

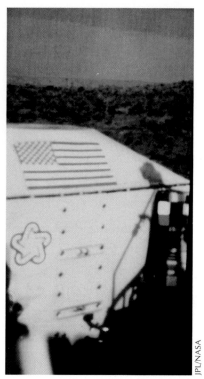

DR. GILBERT V. LEVIN

JPL/NASA

Plate 5. *A segment of the first color image from Mars showing the* Viking Lander *in the image. Note pale blue sky and brownish soil.*

Plate 6. *The orange-red Mars after a JPL technician was ordered to change the colors. This would remain the official NASA colors of Mars until 1996.*

Plate 7. *Another view of the first color image of the Martian surface from* Viking 1 *without the Lander in the photo.*

Plate 8. *The same view "tweaked" by NASA technicians and released to the media as the "true" colors of Mars.*

Plate 9. *This image shows the Viking cameras looking back across the Lan-der. Was this the pre-programmed second panorama that slipped by the NASA technicians and somehow got accidentally published in a NASA exhibit pack-age? Note pale blue sky and yellow-brown soil with greenish areas on and around rocks and soil.*

Plate 10. *The same image "tweaked" to conform to the official NASA orange-red color of Mars.*

Plate 11. *Color-enhanced view of Big Joe boulder at the Viking 1 landing site. Note "cap" of material on top of boulder missing from surrounding rocks and boulders. Also, there is a much greener color covering its surface texture.*

Plate 12.   *Gilbert Levin went on to conduct his own independent study of rocks on Mars. Here, the "Delta Rock Trio" with Levin's highly scrutinized "Patch Rock" (second from left) would display patches of color that varied in intensity and pattern with the Martian seasons.*

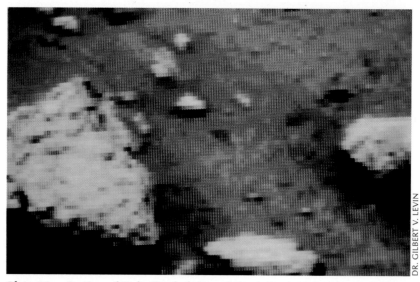

Plate 13.   *Portion of Delta Rock (left) and Patch Rock (right) showing greenish areas.*

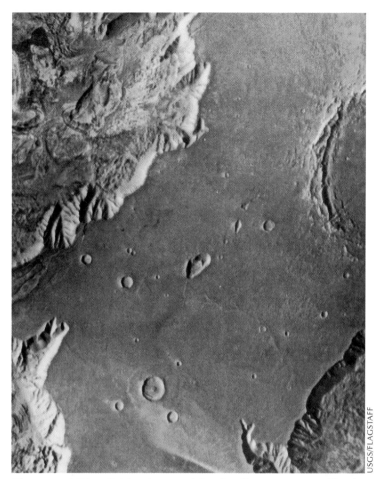

Plate 14. Viking Orbiter *image of Candor Chasm, processed by the USGS to show the complex variation of colors that exist at the surface of Mars. Edwin Strickland III was the first to point out that the Viking 1 landing site had as many as five different soil colors. Levin was first to observe any colors other than orange-red and pointed this out to Strickland and Arvidson.*

Plate 15.   Another USGS-enhanced image with the Viking Orbiter over Claritas Fossae. According to the USGS, this region is "mantled by materials of unknown origin."

Plate 16.   Recently released reprocessed images coming out of JPL/NASA like this Viking 2 image (1996) shows Mars with a blue sky, brown soil, and various rock colors. Is NASA preparing the American public for the images of the surface of Mars that will be taken by the Pathfinder Lander in July of 1997? After twenty-odd years of official "orange-red," will NASA finally admit it has been in error?

JPL/NASA

Plate 17.   *Another of the newly reprocessed* Viking 2 *images. Note the blues and greens going in through the soil and rocks. Algae or similar relatives on Mars? Or something totally alien and unique?*

JPL/NASA

Plates 18 and 19.   *More recently reprocessed* Viking Lander *images now coming out of JPL/NASA.*

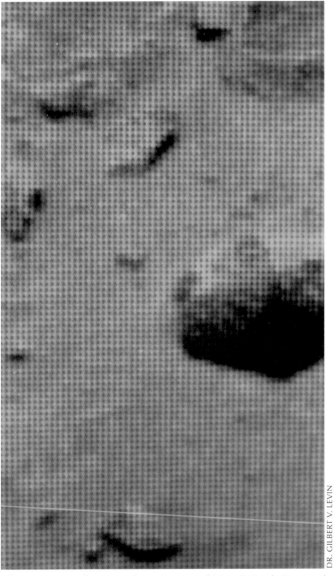

Plate 20.   A Viking Lander 1 *image of "Patch Rock," named by Levin because of the greenish-colored patch.*

Plate 21. *Rock in Antarctica with lichens on it. Compare to "Patch Rock" (Plate 20).*

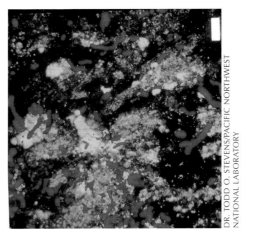

Plate 22. *"SLiMEs!" False-colored digital image of organisms found deep within the Earth that can live without sunlight on just water and basalt rock. The SLiMEs are stained red and the rock on which they are feeding is green. These organisms are approximately the same size as the fossilized bacteria-like structures found in Martian meteorite ALH 84001.*

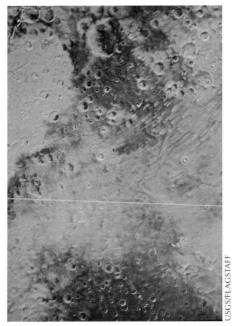

Plate 23. *Viking Orbiter image of blue-green areas that seem to change when the amount of water vapor in the Martian atmosphere is saturated. Most scientists today attribute this unusual display of shifting color to wind-blown dust.*

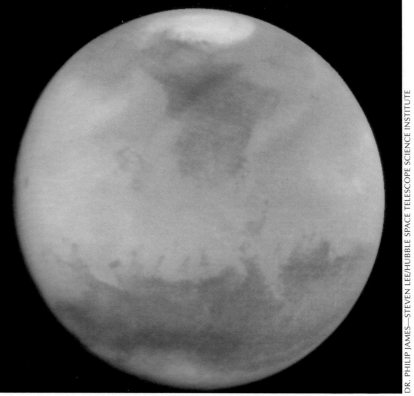

Plate 24.   *A 1995 Hubble Space Telescope photograph of Mars showing its distinctive color patterns. The season on Mars at the time was spring. Were the greenish-blue colors apparent in this image a result of an illusion caused by a contrast effect between the orange deserts and "gray" maria of Mars, as most modern astronomers have asserted? Not in this state-of-the-art photograph!*

CHAPTER SIX

Russian Roulette

It is the policy of this agency that we would rather not have a mission go until we are cleared to establish that there will be no forward and no back contamination, and that is non-negotiable... Maybe we should be sending a sample return mission [to Mars] as the third set instead of the fifth set.

—NASA ADMINISTRATOR DANIEL S. GOLDIN, from the August 7, 1996, NASA Conference Briefing on a Martian meteorite, ALH 84001, found in Antarctica and containing evidence for early life on Mars

Science is the fairest daughter of Truth and Beauty. When politics invades Science, Truth and Beauty perish.

—DR. GILBERT V. LEVIN, 1996

B Y 1992 THE SPACE STUDIES BOARD of the U.S. National Academy of Sciences in association with COSPAR, the Committee on Space Research, recommended changes in the planetary spacecraft sterilization laws, citing a low probability of growth of terrestrial microorganisms on the surface of Mars, and the advantage of saving money. Complete sterilization, called Viking Class Level-2 (a class 100,000 clean-room assembly and component cleaning), costs millions of dollars. The new laws still required any spacecraft carrying a life-detection experiment to undergo a Viking Class Level-2 sterilization, which includes 1) having the life-detection instruments sterilized in a biologically filtered nitrogen atmosphere at 120 degrees Celsius for 54 hours; 2) the entire orbiter and capsule are then heat-sterilized at 116.2 degrees Celsius for 43 hours; 3) an exterior and interior germicidal cleaning of the entire orbiter/lander, including all experiment packages, with alcohol.

The big difference was with spacecraft landers that did not have life-detection experiments on board, because they would now require only the minimum exterior alcohol cleaning. Why? A Viking Class Level-2 sterilization could cost upwards of ten million dollars, depending on the size and complexity of the spacecraft. Concerns about forward contamination of Mars were now relaxed. The only reason landers with biology experiments now required a Viking Class Level-2 sterilization was to prevent the experiments from getting a false positive from terrestrial organisms. Microorganisms from Earth have survived inside the spacecraft on wire bundles, tubing, and other electrical components. They might also survive the space environment, protected by the process of lyophilization, or freeze-drying. This would happen to entrained microorganisms quite naturally on a journey to Mars. The temperature in space is –453 degrees Fahrenheit. Even with a Viking Level-2 sterilization, interior/exterior, thousands of bacteria might survive and be deported on the surface of Mars. The new wave of spacecraft going to Mars with only geology, chemistry, imaging, or seismic instrumentation would only be alcohol-swabbed before launch. The invasion of Mars was about to begin.

Actually, as already mentioned, the early Russian missions to Mars completely ignored any planetary sterilization laws whatsoever. Equally contaminating were the orbiting spacecraft such as *Mariner 9* and the *Viking Orbiters*, which eventually decayed in their orbits and fell to the surface of Mars. Subsurface contamination was thought to be the worst, because some current scientific theories did not consider that microorganisms could survive on the surface of Mars. If the organisms got into the first few meters, however, it was thought they might be able to grow and replicate. Dr. Klaus Biemann, Principal Investigator for the Viking GCMS experiment, made the comment to me in a phone conversation that "If contamination has happened on Mars due to the survival of terrestrial organisms on board early unsterilized Russian spacecraft, it has either occurred very fast within the last twenty-five years, or it may take a period of over a hundred years." Could that mean Earth life can completely overtake and kill all Mars life? No one knows for sure, but it is a possibility. Remember what A. I. Oparin said about one biochemical species encountering another? The one species that was best adapted to the environment would completely overrun any others that would evolve. Likewise, if Martian organisms are dominant in their kingdom, perhaps Earth organisms would not be able to "take over" Mars, and furthermore, if Martian microbes are brought to Earth, depending on the biochemical structure of their makeup, they could perhaps easily overcome our planet. After all, Mars is a much more demanding world than the "comfortable" Earth.

One of the most striking examples of the ability of microbes to survive in extreme environments is the case of *Surveyor III*, an unmanned Moon probe that soft-landed in the eastern shore of Oceanus Procellarum. It touched down on the surface of the Moon on April 20, 1967. It sat in the quiet confines of the harsh lunar environment undisturbed for two and a half years before the crew of *Apollo 12* landed within 300 feet of it on November 20, 1969.

Surveyor III had withstood temperatures of –181 degrees Celsius in lunar night (which lasts about fifteen days) and 101 degrees Celsius in the full glare of the Sun for fifteen days (a half day on the Moon). In addition, the small lunar lander had been subjected to two and a half years of high-energy cosmic ray bombardment, intense solar flare activity, and the vacuum of space.

The *Apollo 12* crew were instructed to carefully remove the *Surveyor III* camera, put it in a sterile bag, and return it to the Earth for examination. The camera was delivered to the Lunar Receiving Laboratory at the Johnson Space Center in Houston, where it was analyzed for any bacterial growth that might have occurred. It was like a scene from the movie *Andromeda Strain*, with scientists and technicians behind a protective glass wall touching the camera only with thick rubber gloves while manipulating it and taking it apart. The team of scientists conducted microbial sampling of a number of different sites on the camera, including the wire bundles. Within about fourteen days' incubation time, a growth started to appear in a culture dish. A species of *Streptococcus mitis* bacteria that had been inside the *Surveyor III* camera, hiding in some foam insulation for two and a half years on the lunar surface, was now growing! How did it survive?

Scientists surmised the organism had been in the spacecraft since its launch in 1967, became lyophilized in space, and had survived in a dormant condition until being revived at the Lunar Receiving Lab. Just what did this incident say about microbes surviving on Mars? Plenty. Mars is a picnic compared to the Moon. It has water vapor, ice, a carbon dioxide atmosphere, and plenty of rocks and soil in which to hide and grow. It was a shocking revelation, although there were some analogous instances on Earth. For instance, the Tomb of Cheops in Egypt was found to contain dormant bacteria at least 3,000 years old. All you had to do was add moisture and viola! They began metabolizing instantly. This demonstrated that microbes could survive the passage of thousands of years. If any microbes had developed on Mars in its early history—and Mars soil is believed to be a mixture of soil and ice—then frozen bacteria or viruses could still be frozen, awaiting our arrival. Is that what happened with Dr. Gilbert Levin's Labeled Release experiment on Mars? Did he, like the scientists in Cheops'

Tomb in Egypt, reawaken sleeping dehydrated microbes with the addition of wet nutrient solution? Or was the soil of Mars alive with microbes even now?

The Russians Want Levin

In March of 1991, Dr. Gilbert V. Levin attended a symposium at the University of Maryland, where a group of Russian scientists and cosmonauts from the Russian Space Research Institute met with their American counterparts to discuss bold new plans of space exploration and cooperation. One of the projects that the Russians had on their agenda was called MARS-94/96, an ambitious effort to return to Mars in 1994 and 1996.

The Russian interest in Mars was straightforward enough. They felt Mars still had traces of a biosphere left, whether contemporary or frozen, and thought that if it did, it could possibly solve the mystery of the origin of life. The main goal for planetary exploration in Russia for the next fifteen to twenty years was going to be Mars. MARS-94 was envisioned as the first stage of this effort that would have an orbiter, a rover, two small landers, and two subsurface penetrators. MARS-96 would be even more ambitious, having an orbiter, a Martian rover, surveillance balloons, penetrators, and small landers. Dr. Gil Levin listened intently to the Russians describing to those gathered at the symposium what amounted to the greatest Mars mission ever conceived. The program only lacked one thing—a life-detection device.

As the symposium continued, the objectives of the MARS-94 Mission were laid out: The Earth-to-Mars flight time would be about 315 days following a Baikonur Cosmodrome launch on a Russian Proton K SL-12 carrier rocket. The MARS-94 Mission was going to be a large international effort as well, including the technology of twenty nations: Russia, Germany, France, Italy, Poland, Spain, Belgium, Austria, Finland, Romania, Norway, Great Britain, Czecho-Slovakia, Ireland, Iceland, Hungary, Bulgaria, Switzerland, Greece, and perhaps the U.S.A.

The instrumentation provided by an international cast sounded no less impressive: ARGUS—Stereo Spectral Imaging System for study of the Martian surface and atmosphere. High-resolution stereoscopic TV camera for detailed study of topographic features with 3D capabilities. OMEGA—visible and infrared mapping spectrometer, for Martian surface and atmosphere. PFS—Planetary Fourier Spectrometer for infrared sensing of the atmosphere and surface. TER-MOSCAN—mapping radiometer to measure the thermal inertia of the Martian soil. Searching for anomalous heat sources and thermal studies of the atmosphere. SVET—high-resolution mapping spectrophotometer, to yield surface composition of rocks and soil and study the nature of aerosols in the Martian atmosphere.

The list of instruments went on and on, but most importantly, MARS-94 would have two small landers with meteorology sensors, color panoramic cameras, seismometer, magnetometer, and alpha particle proton x-ray spectrometer. Perhaps the most innovative scientific instruments carried on this mission would be the two penetrators designed by Finland and Russia to penetrate into the Mars subsurface five to six meters in depth. The penetrators would aerobrake from orbit using a rigid cone, then deploy an air-inflated bag to slow descent to 80 meters per second before impact. Once in and under the soil of Mars the scientific payload on each penetrator would go to work: a TV camera, gamma spectrometer, x-ray spectrometer, neutron spectrometer, and more. In short, the penetrators would provide information on the chemistry of rocks, water content in rocks, physical and mechanical characterisitcs of Martian soil, and imaging of the Martian surface.

The scale and scope of the MARS-94 Mission could hardly be visualized it was so enormous, but there was even more excitement. The landing site chosen for MARS-94 was the Tharsis volcanic region of Mars. Tharsis was the home of the gigantic Martian volcanoes: Arsia Mons, Pavonis Mons, Ascreus Mons, and the largest volcano in the known solar system, Olympus Mons. This was the most interesting site on Mars. Why? A volcanic region meant geothermal activity might still be occurring and observed along with outgassing events such as geysers or eruptions. Not all American scientists thought Mars was volcanically inactive. Dr. Baerbel K. Lucchitta of the U.S. Geological Survey in Flagstaff, Arizona, presented evidence for recent volcanic activity in the nearby (to Tharsis) Valles Marineris equatorial canyon system. He observed in *Viking Orbiter* images what appeared to be dark-gray patches he interpreted as volcanic vents, and because they did not exhibit positive relief, appeared like relatively new formations.

Another American scientist at Lowell Observatory (also in Flagstaff) conducted extensive analysis of the *Viking Orbiter* images and on September 25, 1980, released the following information:

> Dr. Leonard Martin announced today that he has found evidence for an active geyser or steam vent on Mars. A small moving cloud with translucent shadows was discovered on two *Viking* spacecraft pictures. The report continues ... this isolated cloud casts a shadow of unusual shape, showing that it is a slender cone that rises from the surface and narrows to almost a point about one mile nearly straight up, suggesting an eruption. A second picture taken three minutes later shows that both the cloud and its shadow have moved, as well as changing shape. The 400-yard-wide cloud moved about 100 yards to the northeast while expanding upward. These changes are believed to be due to Martian winds. Dr. Martin had been examining the Viking images to correlate information on volcanic activity and Martian winds. The Martian geyser was located just north of Solis Lacus, an area associated with an abundance of atmospheric water vapor.

On December 9, 1980, Dr. Martin again reported new findings of a possible volcanic eruption emanating from a small crater chain south of the Arsia Mons volcano. The official Lowell press release by Dr. Martin reads as follows:

> On August 30, 1977, an unusual small, dense cloud extended upwards from the surface—within the south rift zone of Arsia Mons... Dark radial streaks are seen around most of the larger Martian volcanic mountains and only rarely elsewhere. I suggest a correlation with these ever shifting streaks and volcanic activity.

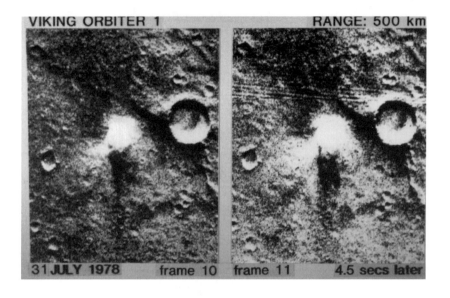

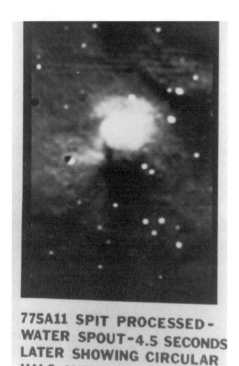

Viking Orbiter *images of a possible geyser or water-spout on Mars.*

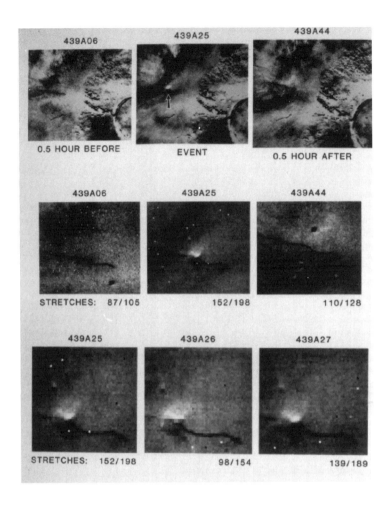

439A06 | 439A25 | 439A44

0.5 HOUR BEFORE | EVENT | 0.5 HOUR AFTER

439A06 | 439A25 | 439A44

STRETCHES: 87/105 | 152/198 | 110/128

439A25 | 439A26 | 439A27

STRETCHES: 152/198 | 98/154 | 139/189

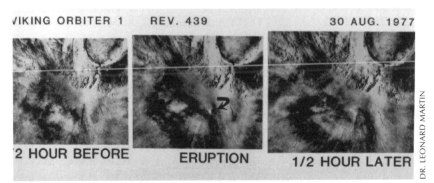

VIKING ORBITER 1   REV. 439   30 AUG. 1977

2 HOUR BEFORE | ERUPTION | 1/2 HOUR LATER

DR. LEONARD MARTIN

Viking Orbiter *images of a probable volcanic event near Arisa Mons.*

213

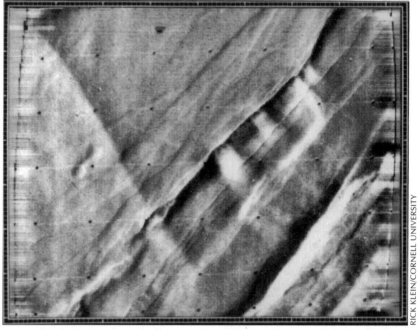

RICK KLEIN/CORNELL UNIVERSITY

Mariner 9 image of an outgassing event coming from a fault-line near the Tharsis volcanic region.

Since Dr. Martin's work on these images, others have been obtained that show interesting features. What was once thought to be nothing more than bright wind-streaks crossing Martian craters are now thought to be frost streaks. How do you get frost or ice crystals going over the top of a crater and being deposited on the opposite side, extending for several times the length of the crater, unless there is some outgassing of water coming from within and being instantly frozen as it hits Martian ambient air temperatures?

Tharsis was perhaps one of the most spectacular areas one could search for life, because even if all the life on Mars had been lyophilized, volcanic activity within the interior along with geysers could thaw it to activity for brief times and most likely longer.

At the University of Maryland symposium, Gil Levin was approached by the Director of the Russian Academy of Sciences—Space Research Institute (IKI RAN in Russian), Academician Albert A. Galeev, along with Professors George Managadze, Roald Sagdeev,

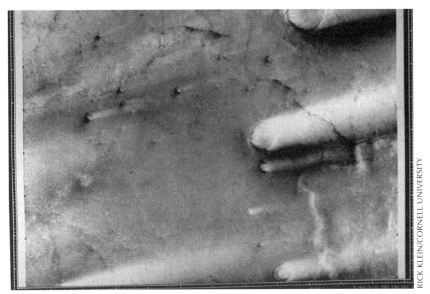

Wind streaks or the newly discovered frost-streaks? 1971 Mariner 9 *image.*

and L. M. Mukhin. All four of these distinguished Russian scientists were aware of Levin's Labeled Release experiment on *Viking* and all of them expressed to Levin that they believed it might well have detected life. In a private informal conversation, Gil Levin was asked if he would be interested in putting the LR aboard a rover on the MARS-94 Mission. Levin jumped at the opportunity, knowing that NASA at the time had no plans to send a spacecraft to the surface of Mars. The one-billion-dollar U.S. *Mars Observer* was in the works, but it was just going to be an orbiter. Levin said he could design a simplified but improved LR instrument capable of resolving once and for all the question of life on Mars. All Levin would need is the approval of his country to do so. At the conclusion of the symposium, Gil Levin told his Russian counterparts they would receive his official proposal in one week.

Levin sent the following letter to Academician Albert A. Galeev, Scientific Leader of the MARS-94 Mission and Director of the Russain Academy of Sciences, on March 25, 1991:

215

Dear Academician Galeev,

Following our discussion at the University of Maryland last week, I have prepared the abstract of a proposal to put a life-detection experiment aboard the MARS-94 Mission. As you pointed out, the time is very short for the incorporation of a new experiment in the '94 mission. However, I have the experiment fully designed and, I believe, its instrumentation would be relatively simple, perhaps utilizing existing components from a spare Viking Labeled Release instrument. The improvement I have designed for the LR experiment should render an unequivocal answer to the life on Mars issue...

Very truly yours,

Gilbert V. Levin, Ph.D.

Levin's improved version of the LR would be the separate application of L and D optical isomers to distinguish between living systems (isomer-sensitive) and chemical or physical agents (isomer-insensitive). One of the unifying themes of terrestrial life is its preference for D-carbohydrates and L-amino acids, including the L-alanine and D-lactate isomers which were in the LR medium. Should both isomers of these compounds yield positive results, the prospect for life similar to terrestrial life would be greatly diminished, and a chemical explanation of the LR results greatly advanced unless contraindicated by various heat control regimes similar to those on *Viking*. Gil Levin did not have to wait long for a response. He received the following fax dated May 6, 1991, from the Director of the Russian Science Institute in Moscow:

Dear Professor Levin,

I appreciate your letter with the proposal to participate in the MARS-94 Project with your LR experiment. Since the implementation of this Project, it has been postponed from 1994 to 1996, and this gives us time to increase the rover payload [now being included] mass and to consider once again the possibility to include the LR experiment in the scientific program of the now MARS-96 mission. [Author's note: the original MARS-96 mission was rescheduled for 1998 when MARS-94 was rescheduled for 1996.] We believe that the total mass of your experiment without the soil sampling system will not be higher than 1 kg. In this case there is a high probability that the LR experiment will withstand the selection procedures successfully and will be included since it is the unique experiment which demonstrated its capabilities within the Viking mission.

From the Space Research Institute, USSR Academy of Sciences, we propose Professor L. M. Mukhin and Professor G. G. Managadze as co-investigators. They've gained much experience in space studies and I think they will help in the matter of our common concern...

Sincerely,

A. Galeev

Levin telefaxed the following response to Professor Galeev on May 14, 1991:

... Your suggestion that Profs. L. M. Mukhin and G. G. Managadze serve as co-investigators is most welcome and I would be pleased to work with them as co-investigator from the U.S. The reason I have not answered earlier is that I have not yet had an opportunity to discuss this matter with NASA to determine its possible support. I hope to do this soon and will then respond in more detail, including specifics concerning instrumentation and the experiment as currently conceived.

Very truly yours,

Gilbert V. Levin, Ph.D.

Gil Levin met with Dr. Carl Pilcher, NASA's Special Assistant for Space Exploration, at NASA Headquarters to discuss the fax he received from Dr. Galeev in Russia. Pilcher presented Levin with a surprisingly discouraging answer: that Levin would have numerous barriers to cross if he wanted to work with the Russians, including clarification with the State Department, the Department of Defense, the Department of Commerce, and others. Not only that, but Pilcher said if NASA were to support a joint-cooperative U.S.-Russian mission to Mars, then NASA would advertise the opportunity to all scientists, which Pilcher stated "could very conceivably leave you off the mission, Gil." Frustrated, Levin replied, "Carl, I have been given a personal invitation by the Russians to represent the United States in an international effort to search for life on Mars. What better way to symbolize our international willingness to cooperate than to conduct the search for life together? The U.S. has no prospect of returning to Mars this century. Whether the answer turns out to be life or chemistry, a major puzzle concerning the surface of Mars will have been solved. My experiment conveys no defense-sensitive hardware, engineering, or scientific information that could conceivably be detrimental to the U.S. NASA support for this project could be quite modest and it would provide an opportunity to resolve the fundamental reason *Viking* spent one billion dollars to go to Mars. There is no downside to this issue. The Soviets have told me that time is running short to include my experiment in the MARS-96 Lander. For God's sake, Carl, the carbon-14 technology used in the LR is forty years old." Pilcher told Levin to have the Russians write to NASA with their request.

Levin approached every U.S. official he could think of who might be able to help him get NASA approval. On November 1, 1991, Levin telefaxed Academician Albert Galeev the following letter:

> Dear Academician Galeev,
>
> Since my last telefax to you, I have sought approval from our Government for me to participate in the cooperative effort you kindly propose. I have met and talked several times with NASA officials including Dr. Carl Pilcher. Dr. Pilcher called me last week following his return from the USSR. He said he saw you briefly, but, unfortunately, did not find the opportunity to discuss this matter. I have also met with Mark Albrecht, Executive Secretary of the National Space Council under Vice President Quayle; Congressman Tom McMillen, member of the Science, Space and Technology Committee; Trip Carey, of his staff; and Dr. Richard Oberman, Science Advisor to the U.S. House of Representatives Subcommittee on Space Science and Applications; and other officials. Last week, I met with Dr. Klein, Viking Biology Team Leader, who, having often said that *Viking* left the Mars-life issue in doubt, expressed interest in the proposed cooperation among Profs. Mukhin and Managadze and me. He promised to help gain consideration for the experiment. All concur that a cooperative effort between our countries to answer the key scientific and

philosophical question, "is life on Mars?", would be highly demonstrative of the improvement in our countries' relationship for the benefit of all.

However, none of these meetings has resulted in an official endorsement of my participation, which I think is very important to the success of the effort, especially if some level of support would accompany it. I am told that I require approval at the Administration level, preferably from the Office of the Vice President, since cooperative space efforts are effected at the "treaty level." It is my hope to see the Vice President to seek his direct approval, but, of course, obtaining the appointment is difficult.

Perhaps a request from an appropriate official in your government could help speed things up...

Very truly yours,
Gilbert V. Levin, Ph.D.

The wheels of scientific progress turn very slowly when politics get in the way. Was it the Mars-life issue itself that was so politically volatile, or was it merely petty indecision on the part of the bureaucracy in government? By January 15, 1992, Levin again received a letter from Dr. Albert Galeev stating that there had been a meeting of the MARS-96 International Scientific-Technical Committee in Moscow, and that it had been decided that if the LR experiment were included on the mission to Mars, it would be called "BIO" unless Levin wanted to change it. Levin telefaxed his nod of approval for the new name "BIO" for his LR on the MARS-96 Mission, but stressed to Dr. Galeev the importance of writing to the Vice President of the United States. On February 5, 1992, Dr. Galeev in a telefax to Levin asked: "Is the level of support from the Vice President of the Russian Academy of Sciences Academician Eugene Velikhov adequate for such kind of business or not? If not, what level of support is adequate?" Levin concurred that a letter from Dr. Velikhov to Vice President Quayle would do nicely. The letter to the U.S. Vice President was drafted and sent. It read:

Dear Mr. Vice President,

As you know, our government is planning a space mission to Mars in 1996. The plan calls for landing a Mars Rover. Last spring several of our space scientists visiting the U.S. met Dr. Gilbert V. Levin, an Experimenter on your 1976 Viking Mission to Mars. Dr. Levin's experiment produced very provocative results indicating the existence of microorganisms on Mars. Because of lack of confirmation by other Viking experiments, this highly important question remains in doubt. However, we have been impressed with his data. Therefore, we extended and Dr. Levin accepted our invitation to include a revised version of his instrument on our MARS-96 Rover. It is important to settle this most tantalizing issue not only in the interest of science, but to help prepare for manned landings on Mars. If there are living microorganisms on Mars, it is vital that we gain knowledge about them to assess suitable precautions for the protection of both planets... Perhaps, if you could spare Dr. Levin half an hour, you would conclude, as we have, that cooperation between the CIS and the US in this project to look for life on Mars in 1996 is in the scientific and public interest.

Sincerely,

Academician Eugene Velikhov, Russian Academy of Sciences

The only reply Dr. Levin got was a phone call from NASA Headquarters' Dr. Carl Pilcher, who told him: "Look, Gil, you need a license to export the LR technology because the carbon-14 radioisotope technology it contains might be classified, so that's it." Levin reiterated his earlier statement about the LR technology being forty years old. It didn't matter. As far as NASA was concerned, the issue was closed. Gil Levin out of curiosity called the Department of Commerce and was told it wasn't NASA that issued those licenses, it was them! The Department of Commerce gave Levin permission to export the LR to Russia with no problem—yet. Levin sent the license to Dr. Carl Pilcher at NASA and waited for a reply.

In the meantime, the new Russia began having problems of its own as changes in the CIS/former USSR had inflation rates soaring. Russia was now facing its most severe financial crisis in recent history. As if things were not tough enough already, Gil Levin received a call on July 29, 1992, from Professor George Managadze, stating that with the current fiscal crisis in Russia, Levin's LR (BIO) would need to be supported financially by NASA if it were to be included on a now-questionable MARS-96 mission. Also, the issue of space-craft sterilization—and its cost—was raised.

Levin sent the following telefax to Professor Managadze on July 30, 1992:

Dear George,

I was very pleased to get your call yesterday. Your report that the manager of the MARS-96 project can accommodate the additional weight we require on the Rover is good news. However, you pointed out the problem of sterilization and asked me to comment on it.

Naturally, a life-detection experiment must be free of contamination by Earth microorganisms. Apparently, the Project Manager fears that the inclusion of our experiment would require "Level-2" treatment of the spacecraft and Rover, that is, heat sterilization. I gather this would impose great difficulty and expense on the mission. I think, however, that it is possible for us to accomplish our purpose without the requirement of sterilizing the Rover and Lander.

Our instrument could be built of sterilized parts in a sterile clean assembly room. The entire instrument, including the nutrients, could then be heat-sterilized and fixed onto the Rover just prior to launch. The instrument would have its own built-in sampling device. The device could consist of a string [like the original "Gulliver"], tape, or flexible notched rod coiled within the instrument. The distal end of the sampler would be inside a one- to several-meters-long sleeve, also flexible. The sampler could be spring-activated, compressed gas-activated, or activated by a small explosive. The area of the instrument surrounding the orifice through which the sampler will be ejected would be covered with a thin shield (plastic or metal). Just prior to ejection of the sampler, the shield would be ejected, possibly by an initial puff of the same pressurized gas ejecting the sampler. The ejection of the shield would leave a sterile surface surrounding the sampler orifice. The flexible tube would be ejected through the orifice to which its proximal end would be restrained by a flange larger than the orifice... The bad news you conveyed to me was that the CIS did not have the money to build the instrument. It had been my understanding since our first discussion that the instrument would be built and paid for by the CIS. This is clearly the major problem we have. The only solution I see is for the CIS to ask the U.S. to participate in the MARS-96 Lander and specifically request support for our experiment. Early this week, I submitted a proposal to NASA to undertake initial studies for an instrument to investigate the LR activity detected on Mars. The proposal sets forth my new ideas on the experiment and instrument. The instrument would have to be very small to be carried by the Discovery and MESUR-type spacecraft NASA is now studying. However, no NASA project for such missions yet exists. The experiment and instrument I proposed are the same I proposed to you and Mukhin for inclusion as our effort on the MARS-96 Rover. I am sending the proposal to you and would appreciate your comment.

Very truly yours,

Gilbert V. Levin, Ph.D.

MOx

Regardless of the reasons NASA did not want Dr. Levin's new version of the LR on the MARS-96 Mission, one thing was clear—it was grounded. Gil Levin, an optimistic person by nature, kept hoping the NASA bureaucrats would recognize the importance of the invitation he had received from Russia.

By August of 1992, NASA had recognized the importance of the MARS-96 Mission and subsequently offered to buy space on the Lander for $1.5 million, but without Dr. Levin or his LR experiment. This action was largely the impetus of NASA's new Administrator, Daniel S. Goldin. The committee, calling themselves the Mars Science Working Group, surfaced again, this time headed by the U.S. Geological Survey's Dr. Michael Carr, who had worked on the Viking Orbiter Imaging Team in 1976. The committee had begun the initial study to place another science instrument on MARS-96 at the same time Levin was trying to get NASA support for his LR on MARS-96. Unfortunately, Dr. Levin had to learn of this development in a newspaper article which stated NASA Administrator Daniel Goldin was buying a place on the MARS-96 Lander. By this time, the Russians had dropped the idea of the Rover on MARS-96 in order to expedite an already highly inflated mission. It was hoped that the Rover could be added to the MARS-98 mission instead.

The new NASA experiment, instead of focusing on the Mars-life issue, was designed to look for oxidants in the soil only. A group of NASA-funded scientists was putting together an experiment and calling themselves the Mars Soil Oxidant Experiment Team. The group's leader was Dr. Christopher McKay, a Research Scientist at the NASA Ames Research Center, considered by many to be one of the foremost experts in the world on living organisms in extreme climates on Earth. Dr. McKay's work included being a Research Diver for the Antarctic Lakes Project in 1986–89. He participated in the Cryptoendolithic Microbial Ecology Research Project with Dr. E. Imre Fried-mann in Antarctica from 1980 to 1989, as well as worked in the Siberian Arctic permafrost regions, the Gobi Desert of Mongolia, the Negev Desert, the Lechuguilla Cave in New Mexico, and the unique fog-cycle life environment of the Atacama Desert in Chile.

Some of the most interesting work McKay and Friedmann had collaborated on was their work on Siberian Arctic permafrost. Ice core-soil samples three million years old were retrieved and studied from various depths in the Earth, and McKay together with

DR. CHRIS MCKAY

NASA's new spokesman for Martian exobiological studies, Dr. Christopher P. McKay, in Antarctica.

Friedmann found that the microbes in the permafrost were still alive after 200 million years of being frozen. The organisms had remained dormant until they were thawed. Then, they immediately began to metabolize. If life were frozen in the Earth for millions of years and still living, then why not Mars? Drs. McKay and Friedmann also found that photosynthesis in certain Antarctic algae remained active at the low temperature of –7 degrees Celsius, a fact that would be important for any existing Mars ecosystem.

SOME OF THE OTHER MEMBERS of the Mars Soil Oxidant group included Dr. Frank J. Grunthaner, senior scientist in the Nanotechnology and Science Group at JPL, and Antonio J. Ricco and Michael A. Butler, senior members of the technical staff in the Microsensor R&D Department at Sandia National Laboratories in New Mexico. In all there were eleven scientists chosen to develop the new experiment. A NASA-Ames public relations expert named Jeff Briggs called Gil Levin to inform him of the project and asked if he would be

DR. E. IMRE FRIEDMANN

Dr. E. Imre Friedmann examines a boulder in Antarctica looking for lichens that grow in rock pores.

interested in becoming a Mars Soil Oxidant Team member. Levin said he would. Briggs said Dr. Chris McKay would be calling him soon. Within the week Levin got a call from McKay, who said that someone at NASA Headquarters wanted him to be included on the team of scientists working on the new project. Levin liked the idea because it would help resolve the reaction he had obtained in his LR experiment. "Would it work?" was the big question, and if it did, would that necessarily mean that just because there are oxidants in the soil of Mars that no life could exist? Levin knew if he did not accept that he might never get back to Mars again with an experiment, so he joined the team thinking he might enhance the Mars Soil Oxidant experiment to provide some new information about oxidants and life.

The first meeting of the Mars Soil Oxidant Experiment Team was at Sandia National Laboratories on October 22, 1992. Levin was there and persuaded the other members on the Team to include a sensor array to look for peroxide in the atmosphere. He pointed out that the oxidant theories all began with the concept that hydrogen peroxide formed in the atmosphere. So, why not look for it there? When encountering the Martian surface, it might be destroyed. All agreed. Levin then suggested changing the name from Mars Soil Oxidant Experiment to Mars Oxidant Experiment, abbreviated MOx, to accommodate its looking in the atmosphere for oxidants. The concept of the MOx experiment was that it would have a broad array of chemicals placed on the ends of fiber optic strands to be probed by light pulses to detect any reactions when the fibers are in contact with the surface. The reactions would be designed to detect chemical oxidants. The MOx experiment would be on each of the two MARS-96 Landers.

In December 1992, MOx was selected as the official U.S. contribution to the Russian MARS-96 Mission scheduled to launch in November of 1996 and land on Mars in September of 1997. Levin received a modest contract from NASA to support his participation as a member of the MOx Science Instrument Team and to do some work in his laboratory to help select and develop thin coatings on MOx. Levin would contribute heavily to MOx. Some of his suggestions that were adapted include: adding a head to sample the atmosphere, applying lead sulfide as one of the coatings because it was the

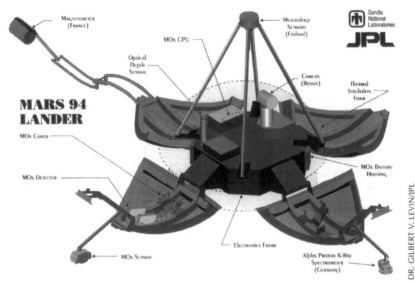

The ill-fated Russian MARS-94 (changed to 96) Lander that carried the JPL
experiment MOx. MOx might have settled the Mars life issue had the mission
proved successful.

most sensitive compound known for the detection of hydrogen peroxide, suggesting a method for breaking the protective seals to expose
the fiber coatings after landing, and incorporating a Trojan Horse.
The Trojan Horse was Levin's covert insertion of a life-detection test
into MOx! Arguing that the amino acid cysteine was a good oxidant
detector, he got it accepted. He then suggested its L- and D- isomers
be applied to separate fibers. Should one isomer respond and not the
other—life! Klein saw through the ruse immediately and said he
would report to NASA Headquarters that Levin was converting MOx
to a life-detection experiment and that spacecraft sterilization would
be required. A Team fight broke out. It was quelled by Chris McKay,
the Team Leader, who, with a wink to Levin, explained that the L-
and D- isomers were just to see if the chemical agent detected had a
chiral preference. Klein still protested, but McKay overruled him
and told him not to make trouble.

Levin became something of a thorn in the side of the MOx Team.
While his technical contributions to the instrument were frequently

accepted, his insistence on keeping the life issue open rankled most. He was quite vocal at Team Meetings in this regard. Soon, the JPL MOx project manager announced that, to conserve costs and increase team efficiency, an Executive Committee of the team was formed. Levin was one of two Team Members not named to the Committee. From that point on, no Team Meetings were held, only Executive Committee Meetings. Levin requested and got a Team Meeting teleconference. By summer of 1996, there were no more communications from MOx. Levin queried McKay and was told that the team had finished its job of readying MOx for launch and had been disbanded. A new team would now be selected by NASA from among those responding to NASA's advertisement.

ALIF

While work progressed on MOx into 1995, Dr. Gilbert Levin was aware of NASA's renewed plan to send "smaller, faster, cheaper" spacecraft to the planet Mars. These included *Mars Pathfinder*, which is basically a test-bed for small micro-rover technology and would not actually include much of a scientific payload. It would have a color camera mounted on the lander and a stereo b&w camera on the microrover. The small 16-kg rover can in essence place its Alpha Proton X-ray Spectrometer sensors up against the side of a rock and analyze the minerals it contains. This will take a period of twenty-four hours to accomplish, through the cold Martian night. Any contribution from *Pathfinder* to the Mars-life issue would be in terms of color images from the surface of Mars. If the cameras were allowed to reveal the true colors and small features of Mars, then perhaps the nature of the green spots on the rocks might be resolved, along with the true colors of the landscape and Martian sky—blue, remember? Launched in November of 1996, *Mars Pathfinder* is tentatively planned to touch down on July 4, 1997.

Then there was NASA's *Mars Global Surveyor*, essentially an orbiter designed to provide detailed views of the Martian surface and to study Martian meteorology. It might also contribute information useful to settling the Mars-life issue if, for example, its state-of-the-art color imaging system showed large areas of green and purple regions changing with the seasons—"the wave of darkening."

Dr. Levin's interest turned to the *Mars Surveyor '98* spacecraft, which would have a lander and orbiter going to Mars on separate launch vehicles in December 1998 and January 1999. NASA scientists are alerted to calls for scientific proposals by means of NASA placing an Announcement of Opportunity in the Federal Register. Dr. Levin saw AO No. 95-OSS-03 for the *Surveyor '98 Lander* payload and submitted his proposal, "ALIF," an acronym for Analysis: Life or Inorganic Forms. While he had been working on MOx, Levin assembled another team of scientists consisting of Dr. Chris McKay as Co-Investigator; Dr. Lee R. Zehner, Biospherics, Inc., Vice President of Science Services; Dr. Ronald Weiner, Professor of Cell and Molecular Biology at the University of Maryland; Dr. Michael R. Belas, Jr., Associate Professor of Marine Microbial Genetics at the University of Maryland; and Sahag Dardarian, an instrument engineer from Swales & Associates Incorporated. Dardarian was a key person in the development of the Hubble Space Telescope's second generation of scientific instruments installed by Space Shuttle astronauts in 1993.

The ALIF goals are outlined in the following scientific abstract submitted to NASA by Levin's team:

> ALIF is proposed for the *'98 Surveyor Lander* as a single instrument to execute an array of experiments aimed at two of the three high-level goals set by NASA. The first goal, "Life," will be addressed through investigating the extraordinary reactivity found in the soil of Mars by the Viking Labeled Release and Gas Exchange life-detection experiments. That activity is generally ascribed to putative oxidants in the Mars soil, but the matter cannot be concluded without direct proof. ALIF will first determine whether the soil at the *'98 Surveyor Lander* site contains the reactive agent found by the two Viking Landers which landed some 4,000 miles apart. Either a positive or a negative response would importantly increase our knowledge of the agent's distribution on Mars. Should the agent be present, an array of experiments

embodying refinements and extensions of the highly sensitive LR radioisotopic technology will seek unequivocal evidence of the chemical or biological origin of the response.

ALIF will address the volatiles and water objectives, another of the Mission's three high-level goals. It will investigate whether reactions detected occur under Mars conditions at ambient water levels, thereby adhering to NASA's desire to focus on water as the common "thread." In further compliance with the Mission's volatiles objectives, the detection of very low levels of organic materials will be sought in the soil.

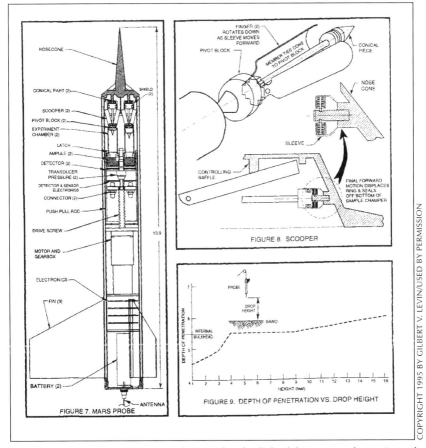

FIGURE 8. SCOOPER

FIGURE 9. DEPTH OF PENETRATION VS. DROP HEIGHT

FIGURE 7. MARS PROBE

One of Dr. Levin's ALIF probes designed to look for life on Mars but rejected by NASA.

The ALIF experiments will be carried out by six pairs (twelve total) of probes launched from the landed spacecraft in ALIF arrows. Each pair consists of an experiment Probe and a comparison control probe. Each Probe will be a self-contained subsystem precluding the possibility of cross-contamination or cross-talk. The first three pairs of Probes are directed toward confirming the Viking LR results and investigating the dependence of this reactivity on temperature, moisture, and stereoisomerism, with an eye toward resolving the chemical or biological nature of the agent. The fourth Probe offers another approach to differentiating between biological and chemical (principally strong oxidants) explanations for the reactivity of the Martian soil. The fifth experiment pair will perform a sensitive test for organics (in particular, amino acids) in the Martian soil as an independent approach to the negative, but very important, findings of *Viking's* gas chromatograph-mass spectrometer. Finally, the sixth Probe will seek to determine the oxidative state of the surface of Mars and will search for activity associated with chemoautotrophic bacteria not detectable by *Viking* and thought to be likely candidates for subsurface life on Mars today.

By ejecting its heat-sterilized Probes beyond the Lander's zone of contamination, ALIF meets the requirements for planetary protection, and for scientific integrity for life detection without incurring the high cost of heat sterilization of the Lander.

Levin sent the proposal off to NASA's Mars Surveyor Program Director, Dr. Joseph M. Boyce, who was also a chief in the NASA Planetary Geology Program Office. NASA responded by mail, citing two reasons it would not accept ALIF: 1) claims that it did not meet the planetary sterilization laws, and 2) it was thought that because the ALIF arrows could hit rocks and be destroyed, projectiles were not satisfactory. Levin thought he might convince NASA to add ALIF to *Surveyor '98*. He called Dr. Boyce and said, "Listen, Joe, even if just one ALIF arrow were to hit its mark—a highly likely situation considering there are twelve arrows—and gather data, it would provide all kinds of useful information." Levin reiterated the fact that the ALIF experiment package would be completely heat-sterilized before being placed on the *Surveyor '98* spacecraft.

Joe Boyce then told Levin it was "really too early in the Surveyor program to look for life" and that ALIF was too large an experiment package. Levin asked, "What if I can truncate it and instead of twelve arrows use only two to look at the L&D isomers?" Boyce told Levin he could submit the idea if he wanted but doubted whether it would be accepted. So Gil Levin called his co-investigator on ALIF, Dr. Chris McKay, who listened carefully to Levin's complaints. McKay suggested coming up with some way of using the UCLA experiments that had already been selected for *Surveyor '98* and offered to send Levin a copy of the winning proposal that had been submitted to NASA by UCLA. McKay sent the proposal to Levin, who spent the weekend looking it over.

Levin called McKay the following Monday and said he thought that the Thermal Evolved Gas Analyzer (TEGA) experiment, contributed to the UCLA package by Dr. William Boynton, might serve their purpose. Dr. Boynton, a Professor of Lunar and Planetary Science from the University of Arizona, was well known in NASA's Planetary Geology Program and contributed many research papers on the composition of the Moon as well as a number of papers on the SNC meteorites, thought to have originated on Mars.

While Levin was trying to get a life-detection device of some type on an American spacecraft, MOx began to turn sour. The program had overspent itself by several million dollars and went through a series of at least seven managers trying to maintain the project on schedule and budget. Worse still was the fact that MOx might not work under actual Mars conditions. Some basic testing had been done with several fiber optic coatings and one or two oxidants. However, MOx seemed to need water vapor to get a reaction. If Mars soil were as dry as some of the Viking scientists suggested, then MOx might not work. Dr. Levin brought this fact up at several MOx meetings and was told that further testing of MOx would cost too much and would jeopardize the device being placed on the MARS-96 Lander in time for launch. Levin said, "What the hell does that matter if the damn thing doesn't work like it's supposed to?"

MOx had questions to resolve on Mars, but in addition to its scientific payload, it was carrying a full cargo of doubts. First, would the spacecraft survive the enormous impact of landing? As I mentioned earlier, the MARS-96 Landers would use descent engines and parachutes, and then four kilometers from the surface, only airbags! They were expected to bounce for twenty minutes, bounding several stories high until dribbling to a halt. Would the petals on the Landers open to initiate the MOx experiment? Even if everything went well, would MOx really work? Only single fibers had been tested, and not under strict Martian conditions. When tests were made, atmospheric moisture seemed necessary for a reaction to take place. That would indicate the reactions take place in an aqueous environment. If so, how does this differ from the living reaction requirement for an aqueous environment? Don't both reactions really have the same fundamental requirements? If so, why was the LR experiment brutalized on the basis of lack of water? And if water is necessary for both types of reactions, how would we know whether chemicals or organisms produced the response—if there is any?

A possible hope lay in Levin's Trojan Horse. If only one cysteine isomer reacted, it would prove the existence of living organisms. If it was the D-cysteine isomer that reacted and not the L-form, the most startling result possible would have occurred—MOx not only would have discovered life on Mars, but would establish that Mars life is fundamentally different from that on Earth. This implies different origins of life—and would open the heavens to speculation. And it all depended on getting back data from four tiny, cysteine-coated fibers, two (L and D) for in the air and two (L and D) in the soil of a planet many millions of miles away. It would now be up to the Russian MARS-96 spacecraft to deliver its scientific cargo, come launch day, November 16, 1996.

232

CHAPTER SEVEN

Last Tango with TEGA

We are a very open agency, and I felt that rather than holding this back in the innards of NASA, we needed to share this with the American people. We need to share this with the world scientific community, that is our first major task.
—DANIEL GOLDIN, NASA ADMINISTRATOR, at the August 7, 1996, NASA Conference Briefing on ALH 84001

Dan Goldin is the only NASA Administrator I have not been allowed to speak with directly since I began to work with the agency in 1959.
—DR. GILBERT V. LEVIN TO BARRY E. DiGREGORIO, 1996

FTER RECEIVING THE WINNING *Surveyor '98 Lander* proposal that Dr. McKay mailed him, Levin spent the weekend studying it. He concluded "My god, the TEGA experiment is a natural," and saw a very easy way to incorporate a portion of his ALIF experiment in it. Levin already was excited about the prospect that the *Surveyor '98 Lander* would touch down at –72 degrees south, right on the boundary of the layered deposits near the southern polar ice cap. The Lander would be launched January 4, 1999, and touch down at the edge of the south polar ice cap on November 27, 1999, late spring on Mars. This season was chosen because it allows the *Surveyor '98 Lander* 92 days with the Sun above the horizon. As late summer and early autumn set in, the temperatures at the south polar regions approach the freezing point of carbon dioxide. The spacecraft is expected to survive until May of the year 2000.

The *Surveyor Lander* features an integrated experiment package called the Mars Volatiles and Climate Surveyor, or MVACS. The MVACS unit has four major elements: 1) a mast-mounted meteorology package with sensors designed to record temperature, wind, and atmospheric pressure; 2) a 6 ½-foot robot arm for sampling the Martian surface; 3) a mast-mounted stereo camera to take stereo photos of the surrounding landscape; and 4) TEGA, or the Thermal and Evolved Gas Analyzer, designed and built by Dr. William Boynton with Dr. David A. Paige from the University of California as a Principal Investigator. The *Surveyor '98 Lander* was the next best U.S. opportunity to have a life-detection experiment, so why didn't it have one, considering that "life" was one of the three goals of NASA's return to Mars?

TEGA was originally planned to have twelve experiment ovens to analyze Martian soil samples for ice content and volatile-bearing minerals. Like *Viking*, a sampler arm (with a small camera attached) is designed to reach out and scoop up some soil and put it into the TEGA experiment, where a sensor will determine the amounts of oxygen, carbon, and hydrogen isotopes contained in the carbon dioxide and water vapor (from ice) gases evolving from the TEGA ovens. Because of volume restrictions, NASA told Boynton and Paige to decrease the number of TEGA ovens from twelve to eight.

Levin thought it was too good an opportunity to pass by. He called Dr. William Boynton to discuss his idea for changing four TEGA ovens into a life-detection experiment, without calling it one. In the conversation Levin told Boynton that he and Dr. Christopher McKay wanted to suggest some "ideas" for him to consider. Levin explained to Boynton that by making small modifications to a few of the TEGA ovens, he could add significant science objectives and enhance the public appeal of the mission. In particular, Levin thought that TEGA should have science objectives that relate to following up on the soil reactivity and absence of organics reported by the GCMS aboard the *Viking Landers.*

In essence, Levin suggested that TEGA could look for the reactive agent(s) in the soil by pre-loading certain compounds into a few of the TEGA oven pans. The experiment would be conducted similarly to the present regime, by adding the soil samples, heating, and monitoring the evolved gases.

Dr. Levin's method to alter TEGA to look for organics, peroxides, and possibly life are outlined in a basic proposal he mailed to Dr. Boynton in Arizona:

> It is proposed to perform the experiment in two to four pairs of the eight pairs of the DSC oven pans in TEGA. Hopefully, only minor software changes, and no changes in the instrument hardware would be required. The individual Viking LR compounds would be used with one substitution: LR's D- and L-alanine would be replaced with D- and L- cysteine because *Viking* found the Martian soil to be very high in sulfur which has been inferred to be in the fully oxidized state... Pre-load two to four pairs of the DSC oven pans in the TEGA. If only 2 pairs, add to pair 1: D-cysteine and L-cysteine, respectively; to pair 2: D-lactate and L-lactate, respectively. If more pairs are available, pair 3:

glycine and formate; pair 4: glycolate and no pre-load (control), respectively. These pans would not be immediately heated to purge temperature upon landing, but would receive soil samples and be subject to 5 degrees Celsius incremental heating at half-hour intervals for up to two Sols....

... Either a positive or negative result for #1 will be of significance in determining the geographical distribution of the reactive soil on Mars. With respect to #2, the compounds will be added to the soil separately (as opposed to the single mixture added in *Viking*), so that the reaction's profile and individual reaction kinetics can help identify the active agent in the soil. Objective #3 will be answered by comparing the respective responses from each of the two chiral compounds. In the case of #4, if each of the compounds exposed to the soil is nearly, or completely, degraded, these demonstrated liabilities would support the hypothesis that soil oxidant(s) are responsible for the reported absence of organics, thereby reinforcing the Viking GCMS result. Finally, objective #5 can be met if the already-planned monitoring of the transition of ice to water is studied for oxygen evolution through the phase change.

Boynton said he thought Levin's enhancement of TEGA was "a great idea" but that pressure by NASA to reduce the TEGA ovens from twelve to eight might not allow for its inclusion. Boynton's experiment was designed in part to look at the possible temperature of decomposition of superoxide/peroxide (if any) in Martian soil. Keep in mind that Boynton served on the Space Studies Board of the National Research Council for a committee called Cooperative Mars Exploration and Sample Return. He voted in favor of a robotic Mars Sample Return Mission directly to the Earth. Boynton suggested that Levin call NASA Headquarters to ask what they thought. Levin called Dr. Joseph Boyce there and requested a meeting with him and Dr. Michael Meyer, NASA's Planetary Protection Officer and Discipline Scientist for its exobiology program.

On February 1, 1996, Levin met with Boyce and Meyer and pointed out that the present plan for TEGA was redundant in that the same experiment would be run on each of the samples of Martian soil placed in the ovens. Levin explained that no changes to the

TEGA instrument would be necessary to incorporate his idea, but a few changes to the software program would be required to accommodate a lower temperature regime. Levin told Boyce and Meyer that the software change would only serve to improve the interpretation of the data from TEGA. Boyce pointed out that the only problem was with the size of the experiment and the budget. Meyer, however, stated that by incorporating Levin's concept into TEGA, it would reduce the original science goals planned for the experiment. He also wondered why Levin selected some of the same compounds used in the Viking Labeled Release experiment. He asked Levin, "What if the same response *Viking* got is repeated?"

Levin explained that his idea was to offer seven compounds separately, thereby determining which of them reacted with the agent in the soil. Meyer countered with the comment that should Levin's enhancement for TEGA be accepted, it would mean that whatever compounds are included in the ovens would be decided on by the TEGA Science Team. Levin agreed and said he had come to NASA Headquarters seeking support for him and Dr. McKay to be included on the TEGA Science Team.

Boyce then posed the question of whether the *Surveyor '98 Lander* would have to undergo the now extremely costly Viking Level-2 sterilization to accommodate Levin's enhancement of TEGA, because as pointed out in Chapter Six, all spacecraft carrying life-detection experiments have to submit to the Level-2 sterilization procedure. Levin smiled and said, "This is not a life-detection experiment." He further explained that his enhancement of TEGA might, however, "influence" the final interpretation of the Viking LR results.

The meeting then turned to the subject of spacecraft sterilization. Levin mentioned that it was odd NASA held the position that it should heat-sterilize spacecraft only with life-detection experiments and not those with geology, chemistry, or meteorological experiments. If NASA truly felt that terrestrial microorganisms could not survive on Mars as Mike Meyer reiterated to Levin, why sterilize at all? Paradoxically, NASA's position on this issue was that "life-detection experiments" might get false positives from the terrestrial microorganisms surviving inside the spacecraft, so therefore the costly

Level-2 sterilization was necessary! There was obviously a conflict here.

Meyer explained that since the *Surveyor '98 Lander* experimental regimen was confined to the surface and would not be digging deep holes where Martian microorganisms might exist or survive, perhaps the TEGA enhancement suggested by Levin would not cause the sterilization issue to arise. After all, NASA's position since 1991 had been that no terrestrial microorganisms could "grow" within the first few centimeters of the surface based on the assumption that strong organic molecule- destroying oxidants were in the soil. No one at NASA considered that the oxidant theory could be wrong! Yet, other problems lay ahead for Levin.

At the conclusion of the meeting, both Boyce and Meyer felt it was essential for Levin to get Dr. William Boynton to formally request the enhancement for TEGA since it was his experiment. Levin said he would contact Boynton and left.

Levin went to see Boynton in March of 1996 at the University of Arizona, where they could discuss in detail the aspects of altering TEGA to a partial life-detection instrument. Boynton said, "That's a very exciting idea, Gil. I have been hoping to pay attention to this area of science and we do have an oxygen sensor on board for this purpose of detecting oxidants ... but I had not thought of the amino acids that you suggest as a means for seeking life." Boynton then said his only real plan to address the Viking results was to look at the decomposition of peroxide or superoxides in the soil, but he was very taken by the idea and would discuss the issue with Chris McKay. Gil Levin left Boynton's office in high spirits and returned to Biospherics, Inc., in Maryland.

Shortly thereafter a *Surveyor '98 Lander* meeting was held at the Jet Propulsion Laboratory in California with Boynton, Paige, McKay, and others, but not Gil Levin. After the JPL meeting McKay sent Levin a message that summarized the conversation. He seemed encouraged that Levin might be included on the Surveyor '98 Science Team, but then explained that Boynton and Paige felt uncomfortable about looking for life. Levin expressed his disappointment at not being included in the discussions and asked McKay why there was a problem with seeking life since it was a NASA stated goal. It was the same "life-shy" attitude that NASA had adopted since Viking, essentially convincing themselves that it could not exist presently and refusing to consider otherwise.

Levin knew this routine all too well and responded in the only way that he would be allowed—to design the TEGA enhancement with a slant towards looking for oxidants. Levin wrote to Boynton stating that he thought it was important to seek the "putative" oxidants and agreed to spiking the TEGA ovens for that purpose. Levin proceeded to tell Boynton that the use of L- and D-cysteine was included on the MOx experiment (Levin's Trojan Horse on MARS-96), which could provide interesting clues to the nature of oxidants if present in the Martian soil at the south polar cap *Surveyor '98* landing site. Levin also suggested the use of sulfate in one of the TEGA ovens to provide a test of whether any reducing activity may exist which might indicate redox cycling. Levin was now hoping for only three of TEGA's eight ovens and requested that if Boynton and Paige decided not to include him on the Science Team that his TEGA enhancement concepts be kept confidential and not be put to use.

Things once again did not seem right to Levin. Boynton, who had initially said he considered Levin's idea to be excellent, was now telling him that he did not think it could be placed on *Surveyor '98*. What had suddenly changed his mind? In the meantime, in the May 22, 1996, edition of the publication *Investors Business Daily*, a front-page article boldly quoted NASA Administrator Daniel S. Goldin: "NASA wants to learn whether life is on other planets, not little green men, but single-celled life." This same theme was being reiterated over and over again to the media, and yet privately NASA would not allow a life-detection experiment to fly on board any of its new missions to Mars. Fortunately, the JPL MOx experiment had Levin's "Trojan Horse" en route to Mars aboard the Russian MARS-96 spacecraft. This now seemed the only hope for settling the issue of current life on Mars, and it had to be done by calling it an oxidant experiment!

Viking Anniversary 1996

Invitations had been sent out by May of 1996 for the second ten-year gathering of the Viking Alumni. Dr. Gilbert V. Levin did not receive one and decided to call NASA about it. They said it was in the mail and he should get it shortly. It never came. However, Levin made reservations anyway for the big event being held July 20 at the National Academy of Sciences auditorium in Washington, D.C., featuring a number of guest speakers followed by a dinner for former Viking scientists and staff.

That hot July day, Gil Levin sat among some of his Viking peers who had worked with him on the historic mission twenty years earlier. Dr. Norman Horowitz was not there. Vance Oyama was not present; he had quit NASA-Ames shortly after Viking and no one from the Viking program has heard from him since. Dr. Klaus Biemann, Principal Investigator for the Viking GCMS, was in attendance. Dr. Thomas Young, former Viking Mission Director, gave a brief speech on the scientific rewards of the Viking program. He then turned the podium over to Dr. Michael Carr, who worked on the Viking Imaging Team and was a geologist from the U.S. Geological Survey's Astro-Geology Branch. Carr presented a geological history and evolution of Mars. The program continued with Donna Shirley, JPL Mars Exploration Manager, who spoke on NASA's near-term Mars exploration plans.

Then, Dr. Christopher McKay presented some slides of Mars and said, surprising no one, "The mission of Viking was biology and the search for life." He further commented that Viking had three biology experiments on it. McKay said to his audience that the current popular consensus was that no life was detected on Viking. Later he stated that life might have been detected by Viking and that the search for life should continue because life still might be found in isolated "oases." McKay then presented some images of Mars taken by the *Viking Lander* cameras. One image, a slide of "Big Joe" boulder at the *Viking 1* landing site, stuck out like a sore thumb. It showed Big Joe with green areas all over it, and no one including Levin made any remarks about it.

By noon, McKay had concluded his presentation and NASA Administrator Daniel S. Goldin took the stage and proceeded to talk about current NASA plans for Mars exploration. He spoke in a politically correct manner, telling his audience how NASA is going to first look at the geology of Mars, then go look for water, then measure the environment, and finally send a robotic Mars sample return mission. Goldin knew one thing his guests did not: he had been briefed on the Martian meteorite found in Antarctica (ALH 84001) containing possible fossil bacteria from Mars. He smiled confidently after each and every sentence.

At the conclusion of Goldin's speech, panel moderator Donna Shirley opened the meeting to questions from the audience. For whatever reason, Levin chose to remain silent, playing the part of the observer this time. He listened carefully to every word. Lines of people wanting to ask questions extended toward the rear of the auditorium. Some of them would be in line for more than half an hour. Nearly last in line was Dr. Klaus Biemann, the Viking GCMS Experimenter. When it was Biemann's turn to ask a question, he instead started to give a speech on how much he disagreed with the plan for Mars exploration Daniel Goldin had just presented. Biemann disagreed with Goldin's view of sending humans to Mars, one of Goldin's main themes. Donna Shirley, who obviously did not recognize Dr. Biemann, interrupted him: "Sir, are you going to make a speech or ask a question?" Many in the auditorium then began laughing, but Dr. Biemann, seemingly unaffected, persisted in asking, "Why are we looking at Mars exploration this way?" Daniel Goldin, also not recognizing Biemann, responded with a smile, "Sir, there are very good reasons," but Biemann interrupted with, "No, there aren't! We already know everything we need to know about Mars and if there is anything else we can get it by robots." Daniel Goldin then turned on a serious expression and said, "Sir, I think you are being cynical. I have been very open with you, why are you not being open with me?" Biemann returned to his seat.

Later that night, after the Viking anniversary dinner, Gil Levin drove back to Maryland. There was business to attend to at Biospherics and the impending TEGA matter. Levin had not heard any encouraging news as of yet, so he decided the next day to go to the top. He thought he would try to get through to the NASA Administrator himself, Daniel S. Goldin. His call was forwarded to Dr. Jurgen Rahe, Head of the Solar System Exploration Division at NASA Headquarters. Rahe said he was on another call and would call Levin back, and within a few moments, did so.

Levin recalls, "I wanted to appeal to Dan Goldin to support my TEGA idea. Also, since I had met with every NASA Administrator except him, I thought I would like to meet him." However, Rahe, perceiving an end-run, said, "Look, there's no way you are going to talk with Goldin about getting any money—he's not going to give you any money at all." Levin said, "Okay, I won't talk about money, I just want to talk to him!" Rahe, who had previously been helpful within the bureaucratic bounds of NASA, explained that all business pertaining to *Surveyor '98* would go through him. Levin did not get an audience with Goldin. Levin wrote and e-mailed Goldin but received no response.

On August 5, 1996, Gil Levin picked up a copy of the periodical *Space News*, which leaked the story that NASA Johnson Space Center scientists had found evidence for early life on Mars. The article stated that an official NASA announcement would coincide with the scientists publishing their results in the September 16 issue of *Science*.

The story was too big to wait until September 16, and someone had leaked key information. As Levin was riding home from work on August 6, 1996, he heard on his car radio that CBS planned to break the story on the evening news that night—and they did. On August 7, 1996, NASA made the announcement to the world that evidence of fossil bacteria and organic material had been found in one of the class of SNC meteorites (covered in the next chapter) that were believed to have come from Mars. President Clinton himself preceded the Press Conference, acknowledging the discovery as one of the most important in scientific history, if true. Gil Levin could not have been happier—he thought for sure NASA would have to look at his TEGA idea more closely now. After all, he had just heard President Clinton himself say, "I am determined that the American space program will put its full intellectual power and technological prowess behind the search for further evidence of life on Mars...." The JSC NASA scientists had studied the meteorite known as ALH 84001 for two and a half years. They found organic material, carbonate inclusions, magnetite, and strange-looking tubular structures thought to be fossilized bacteria.

The week following the NASA Press Conference on ALH 84001 had the entire NASA infrastructure buzzing with activity, and Levin decided that considering the circumstances, he would again try to spark NASA's interest in putting a modified life-detection experiment aboard TEGA, virtually costing NASA nothing. Levin sent Dr. William Boynton an e-mail pointing out that six months had elapsed since the idea was first proposed. His intent was to make a case for reconsideration, citing that the recent report for possible evidence for early Martian life had re-energized the search for life on Mars. Levin reminded Boynton that the earliest opportunity for the U.S. to look for extant life on Mars was the *Surveyor '98 Lander.* There was no interference with the original goals of TEGA and no compromising of the planetary protection sterilization requirement. Levin thought Boynton would now jump at the opportunity to solve the Mars-life issue.

Levin stated: "I believe that if we miss this opportunity, someday we will have a difficult time explaining to inquiring officials and the public. Sure, it will take more work and impose some inconveniences, but isn't that what the President, the NASA Administrator, and the public expect of us?"

Dr. Boynton e-mailed Levin a final response, saying he feared the TEGA experiment with Levin's enhancements would give a false positive for life on Mars because the *Surveyor '98 Lander* was not going to be given a Level-2 sterilization, and that meant at least some Earth microorganisms could survive and find their way into Levin's experiment. No, unless NASA was going to spend the millions of dollars necessary for sterilizing the *Surveyor '98 Lander* in and out, Boynton felt it was no use putting a life-detection experiment aboard.

On September 23, 1996, Gil Levin as a last measure sent Boynton another e-mail providing mathematical calculations describing all the potential terrestrial organisms that might survive inside the

Lander. He concluded that their numbers would be so few, it would not interfere with differentiating the organisms that might be present in the Mars soil. Levin waited but got no response from Boynton. It seemed the message was clear. No life-detection instruments were going on *Surveyor '98*.

All was quiet until the first week of October, when Gil Levin began to read the latest issue of *Space News*. His jaw dropped in utter disbelief. An article reported that plans were being made to modify the TEGA experiment. Despite the fact that Levin had first suggested modifying TEGA, he was not being included.

Levin immediately called Dr. Jurgen Rahe for an explanation. Rahe said, "Gil, the answer to your TEGA proposal is in the mail." Levin then called Joe Boyce at NASA: "Joe, is the answer in the mail from Rahe a rejection?" he asked. Boyce answered, "Yeah, probably." Levin exploded, "This has blown it, Joe! I know you have tried to help me, but I am not going to let this happen. NASA is not going to take my idea and run off with it!"

Joe Boyce explained, "You said you didn't want to do just the chemical aspect of it, right?" Levin said, "That's correct, not just the chemical aspect." Boyce then proceeded to tell Levin that the Surveyor scientists thought it was okay for them to use the idea for chemical purposes. Levin then replied, "I don't give a damn whether they want to do it, it's my idea and they are not going to steal it!" Joe Boyce told Levin to calm down and that Dr. Boynton and his TEGA Team were in France for a meeting. "Gil, don't do anything until they get back on Friday. I'll see what I can find out."

Levin waited for Boyce's phone call that Friday. The phone rang and Levin answered. Boyce said the reason Levin's TEGA proposal didn't go through was because of spacecraft sterilization problems. Levin said to Boyce, "You said NASA approved my method to protect the spacecraft against a false positive without requiring sterilization and now you are retracting it?" Joe Boyce calmly stated, "We didn't retract it, Gil, JPL just said it had to be proven with an engineering study and we calculated it would cost $25,000." Levin, completely surprised, then asked, "You mean to tell me that for $25,000, NASA is not going to look for life on Mars when the next opportunity isn't for ten years?! I find that completely ludicrous!" Joe Boyce responded, "We have a tight budget, Gil, I'm sorry." *(See Color Plate 20.)*

Levin then told Boyce, "NASA made sure I couldn't put the LR on the Russian MARS-96 Mission even though I was given an invitation by the Russians to participate in a search for life on Mars. NASA has, in effect, kicked me off the MOx Team after I contributed very heavily to its design and science! NASA has been denying my results on Viking for years, and now has stolen my idea for this latest experiment on TEGA. That's going too far, Joe!" *(See Color Plate 21.)*

The phones hung up and Gil Levin was now out of the loop— the scientist who had originally suggested to the first NASA Administrator in 1959 that a search for life on other planets should be implemented, one of the first NASA exobiologists whose "Gulliver" life-detection device was the centerpiece of the NASA Exobiology Department for years, who had worked with *Mariner 9* in 1971 and *Viking* in 1976, and who had fought like hell for more than twenty years to resolve one of the most important issues facing the human race. It wasn't going unnoticed, however. Levin himself wrote an editorial for *Space News* in its October 14–20, 1996, issue that questioned NASA's reluctance to look for life on Mars. Levin pointed out in the editorial that he had developed a virtually cost-free method for NASA to look for life on Mars using his idea to enhance the *Surveyor '98 Lander* TEGA experiment.

The Boston Globe took an interest in Levin's story and began to write several articles on NASA's reluctance to search for life on Mars. Soon afterword, *The Boston Globe* put the articles on its Web Site. Next, *The London Times* ran a story on November 11, 1996, titled "The Secret of Mars," in which Levin's discovery of possible life on Mars in 1976 was outlined. The Annapolis *Capital* followed with a similar story. The heat was intensifying.

However, fate would deal a cruel blow to Gil Levin and all of us who want to know about life on Mars. On November 16, Levin and his wife Karen were sailing in their boat on Chesapeake Bay, relaxing and enjoying the day. Levin had just listened to the news of the

Russian MARS-96 launch on his radio. Everything was fine, it seemed. His chemically modified optical fibers on the MOx experiment (the Trojan Horse) would prove once and for all whether oxidants caused the reaction in the LR in 1976, or whether it was Martian life. Levin couldn't have been happier that day, at that moment.

As evening approached, Gil and Karen Levin sailed to shore and went to a social gathering. There a friend broke the terrible news: the Russian MARS-96 spacecraft launched earlier in the day failed to achieve Earth orbit due to a malfunction preventing the fourth-stage booster from igniting. Nearly twenty-four hours later, the MARS-96 spacecraft, representing an international cooperation of more than twenty countries, and which was the last hope in this century of settling the Mars-life issue, reentered Earth's atmosphere in a ball of fire, scattering debris over a 200-mile-long portion of the Pacific Ocean, Chile, and Bolivia. Eyewitnesses in Chile reported seeing a comet-like meteorite breaking apart over the Atacama Desert.

Both the Russians and the U.S. Space Command Space Surveillance Network tried tracking the incoming MARS-96 probe but lost its precise position. President Clinton made a diplomatic phone call to Australia to warn them of the potential impact. Apparently, the U.S. Space Command could not pinpoint the exact location MARS-96 would come down. There was reason for alarm. The MARS-96 probe contained four small batteries containing 200 grams of plutonium 238. Plutonium is considered to be one of the most dangerous and toxic substances known. Just one-half gram of plutonium 238 in any groundwater supplies could kill thousands of people if ingested.

The Bolivian and Chilean government officials were outraged that they did not receive a similar warning from the United States as did the Australians. The issue at the time of this writing is volatile — with both of these South American countries wondering if the lives of Australians are worth more than those who live in Bolivia and Chile.

Levin was devastated. He had managed to get on another Mars mission, no easy task for any scientist, then have it all end this way. The only hope of looking for life on Mars now would be with the *Surveyor '98 Lander*, using Levin's enhancement of the TEGA experiment. That now seemed even less likely to ever happen. NASA was more interested in looking for chemicals in the soil—chemicals which, even if they are present, do not exclude the possibility of microorganisms living in and under the soil surface of Mars.

Why Not?

What possible reasons could NASA have to delay the search for life on Mars, though it is stated time and again as one of its three main goals for Mars exploration? Unfortunately, to try and understand this problem we must rely to a degree on speculation. Let's examine four possible scenarios:

1. NASA WANTS EVERYONE TO BELIEVE Mars is a dead, sterile world in order to clear the way for a Mars Sample Return Mission. This is a serious matter. If this scenario were correct, it would likely mean that NASA is not in control; that instead the Department of Defense has taken over. Why bring back a potentially pathogenic virus or bacteria directly to the Earth, risking the destruction of our biosphere? A calculated risk? Or is it the same risk involved in developing new biological and nuclear weapons? Is this the reason? Whoever gets to study a Martian virus or bacteria might develop an antidote first? This type of situation would require security at the highest level, similar to the measures taken with nuclear weapons facilities.

If you recall how H. G. Wells' classic science fiction novel *War of the Worlds* ended, with the Martian invaders being destroyed by terrestrial bacteria and microbes (the common cold!), it illustrates vividly what could happen if you reverse that situation with harmful organisms brought back from Mars! Historical reminders of how dangerous exotic organisms can be include the following: During the 1300s, about one quarter of the European population died as a result of Asian bacteria—the plague. Smallpox brought to the Americas by the Spaniards caused the death of untold thousands of native inhabitants. The smallpox mortality rate was once as high as

95 percent. The first contact with the Europeans proved deadly for the native peoples inhabiting the Polynesian and Hawaiian Islands in 1778, with over half of the native population wiped out within a fifty-year period. In 1875, the king of Fiji, returning from a short voyage to Australia, brought back with him measles, which killed more than 40,000 of his people.

It is not just human pathogens we should worry about, either. Consider what would happen if invading Martian organisms found sea plankton palatable. It would affect the entire food chain and mean a total reduction in oxygen production, perhaps converting our atmosphere into something more like home—carbon dioxide! Maybe a Martian pathogenic organism would wreak havoc on world grain supplies, ruining crops around the globe. The point is, we just don't know enough about Mars yet to justify bringing a soil sample back to analyze. It is just too risky. Those who are reluctant to learn from history are doomed to repeat it, as the saying goes. Every citizen on Earth should absolutely refuse to support any space program that would "chance" such a potentially deadly threat of a premature Mars Sample Return Mission. Biology experiments need to return to Mars and study it for years, not just for one or two missions!

But as of this writing NASA has literally refused to put another life-detection experiment on its new era of spacecraft, fearing that because they can no longer afford to sterilize their spacecraft, looking for life is too expensive! And they now are going to do something terrible, that at the time of Viking was considered unthinkable—the willful biological contamination of Mars by unsterilized NASA spacecraft landers. Looking for life on Mars cannot stop with just a simple examination of the rocks and surface. No, this must include looking for lyophilized organisms hiding in subsurface permafrost regions, and possible extreme halophiles that can live for millions of years inside salt crystals. NASA has been side-stepping the issue of potential biological contamination of the Earth by Martian organisms because they know how long such biological research may take, which is ten to twenty years by robots. Every time NASA mentions how soon they would like to bring back a Mars sample, everyone who cares about the Earth should call NASA Headquarters in Washington,

D.C., and ask them what they are doing about resolving the Mars-life issue, and then ask why NASA's goal is to bring back a sample of Mars before they know for sure whether it is alive and pathogenic. Ideally, as awareness of this issue spreads, the Department of Agriculture, the U.S. Public Health Service, the Environmental Protection Agency, the World Health Organization, and other groups will keep a watchful eye.

It is interesting to note that NASA's Dr. Wesley Huntress was asked a relevant question at the August 7, 1996, Press Conference by *Boston Globe* reporter David Chandler:

> One of the three principal investigators of the Viking life-detection tests has maintained steadfastly for the twenty years since that experiment that his Labeled Release experiment did in fact produce evidence of current life on Mars. That claim has not received a lot of respect in the scientific community. In light of this evidence of past life on Mars, do you think there will be a re-examination of the results of that experiment and perhaps a more receptive response to his proposal for a follow-up experiment that would resolve clearly the results of that experiment, one way or the other?

Dr. Huntress responded,

> This result that you heard about today (ALH 84001) has only something to say about the possibility of life early in the planet's history, not really about the existence of life of any sort on the planet today. If in fact through the scientific process, if it is determined that in fact this is good evidence of early life on Mars, then that is really all it is, but it would in fact raise the possibility that life may have continued to evolve on the planet and I will ensure at that point, the scientific community would pass some judgment on whether or not we should re-open the issue of the Labeled Release experiment on the *Viking Lander.*

Perhaps this reassuring-sounding statement from NASA is the only guarantee we have. After all, what responsible scientist would put his own world at risk? Only a modern-day Nazi regime would do such a thing without consulting its citizens, right?

2. ARE NASA AND THE UNITED States Government afraid of the consequences that the announcement of current life on Mars might have on a religious or cultural level? Dr. Donald N. Michael and seven other collaborators in a summary report filed in December of 1960 entitled "Proposed Studies on the Implementation of Peaceful Space Activities for Human Affairs," prepared for the Committee on Long-Range Studies of the National Aeronautics and Space Administration, had this to say:

> The Fundamentalist (and anti-science) sects are growing apace around the world and, as missionary enterprises, may have schools and a good deal of literature attached to them. One of the important things is that, where they are active, they appeal to the illiterate and semi-literate (including, as missions, the preachers as well as the congregation) and can pile up a very influential following in terms of numbers. For them, the discovery of other life —rather than any other space product—would be electrifying... If plant life or some subhuman intelligence were found on Mars or Venus, for example, there is on the face of it no good reason to suppose these discoveries, after the original novelty had been exploited to the fullest and worn off, would result in substantial changes in perspectives or philosophy in large parts of the American public....

While the Brookings Report played down the potential impact of discovering simple one-celled organisms on other planets, it did point out that the discovery of "intelligent life could cause a world-wide crisis." Just how people would react to the announcement of alien life was realized in part August 7, 1996, when NASA told the world that possible evidence of ancient life on Mars was discovered in a meteorite from Antarctica (ALH 84001). There were the anticipated skeptics, but surprisingly, it seemed business as usual. Society in America had just passed its first life-on-other-worlds litmus test. However, this was only evidence of microorganisms that may have existed billions of years ago. What if society knew there were still living organisms on Mars today? What if Panspermia is proven true? What if we discover by examining life on Mars that life has no beginning—that it has always co-existed with inanimate matter? What are the social implications of discovering that the universe may never have had a beginning and may consist of countless other universes? Such concerns are exactly what NASA's new *Origins* program is all about. At the request of the White House Office of Science and Technology Policy and the National Research Council, dozens of biologists, planetary scientists, astronomers, cosmologists, and theologians met for a series of in-depth discussions about the implications of future discovery and the possibility of wide-spread life throughout the universe. Vice President Al Gore hosted the sessions, which took place October 28–30, 1996. NASA Administrator Dan Goldin later voiced concerns about religious fundamentalists who might refuse to support NASA funding if discoveries are made that offend or threaten their belief systems.

However, if belief-shattering truths are discovered by scientists studying the solar system and the cosmos, then we must accept them, or otherwise be doomed to mediocrity and societal stagnation. To ignore the truth about nature is to deny truth about God. Spiritual and social harmony must be based on truth—not myth. It is now the higher responsibility of religious leaders to pave the road to the revelations about to unfold by working hand-in-hand with scientists and educators. This is a major challenge in the twenty-first century.

3. NASA'S BUDGET IS UNDER ATTACK. Almost all of its budget goes to the manned space program, with little left over for true science, and NASA wants to protect that. They have announced a program of Mars exploration that in their view is a very orderly program starting with (again) going back to look at the rocks and what they are made of; then looking at the geology and chemistry of the environment, the temperature and humidity —essentially hole-plugging and repeating what the *Viking Landers* did. Also falling under this category of budget-slashing is sterilization, because of the enormous costs. However, considering that NASA has readily accepted contaminating Mars with "exterior only" sterilized spacecraft that do not have any life-detection experiments, it seems irregular that an agency that professes to consider looking for life a high priority could not find a cost-effective way to do so. Dr. Levin suggested an inexpensive solution

for TEGA and was rejected. So it doesn't appear to be just a fiscal issue.

4. NASA HAS CHANGED ITS NEGATIVE presumption that life cannot presently survive on Mars and is preparing to announce it, if the *Pathfinder* or *Global Surveyor* camera systems show evidence for it in 1997. This evidence would be in the form of unique colorations in the soil and rock surfaces that change with the seasons on Mars. Also, views from orbit confirming a global color change from season to season would be evidence of the age-old wave of darkening theory. But what of Drs. Levin and Straat, who discovered evidence for extant life on Mars in 1976? Surely at this time NASA will convene a meeting and appropriately congratulate these pioneers.

One of the many surprises in 1996 aside from ALH 84001 was the October 24 announcement of Pope John Paul II on evolution: "... new knowledge leads to recognition of the theory of evolution as more than a hypothesis." What new knowledge was the Pope referring to? Was he influenced by the August 7, 1996, NASA announcement of past life on Mars? Perhaps the Pope is well aware of the changes in store for the next age of discovery —that life is abundant in the cosmos—thus paving the way for a new human spirituality in the year 2000. Four years earlier the Pope declared that the Church erred in condemning Galileo Galilei as a heretic in 1633 for contending that the Earth was not the center of the universe. The significance of these events is not trivial and seems to indicate that much of global society is ready to cast away the chains of a bio-centric Earth.

The SETI program (Search for Extraterrestrial Intelligence) has been searching the depths of our galaxy for years, seeking evidence of intelligent life "out there" via radio telescope frequencies. They hold out the hope that an advanced civilization might use a similar technology. Let us hope that the way NASA has handled the Viking LR results is not the reaction imposed when a radio telescope finally picks up a signal from another civilization out in space. If a signal is picked up within the next decade, would NASA scientists, as they did in the case of Viking, then try to explain it in terms of anomalous signals coming from the Earth's atmosphere? Or would they come up with exotic theories of effects caused by solar flares that can never be proven? Perhaps it would be only after billions of dollars and years of eliminating every other possibility that NASA scientists finally announce, "Yes, it is a signal from space!"

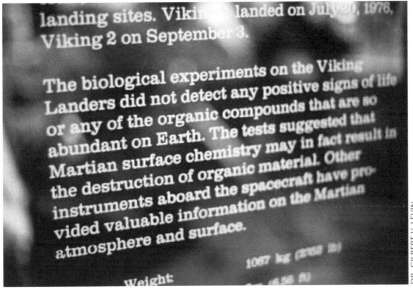

landing sites. Vikii landed on July 20, 1976, Viking 2 on September 3.

The biological experiments on the Viking Landers did not detect any positive signs of life or any of the organic compounds that are so abundant on Earth. The tests suggested that Martian surface chemistry may in fact result in the destruction of organic material. Other instruments aboard the spacecraft have provided valuable information on the Martian atmosphere and surface.

Weight

1067 kg (2352 lb)

The plaque placed in front of the Viking Lander Exhibit at the Smithsonian National Air & Space Museum in Washington, D.C., declaring "no signs of life" were detected by the biology experiments.

CHAPTER EIGHT

ALH 84001 and a Martian Microbial Ecosystem

There are fingerprints of Mars all over these (SNC) meteorites....
—Dr. Michael Carr, Ph.D., former Viking Imaging Team
Geologist, 1996

I n 1969 a scattering of meteorites was found on the surface of the ice in Antarctica. Because there is little precipitation in the form of snow in Antarctica, the meteorites have remained exposed for thousands of years. The dark crust of these meteorites contrasts with the surrounding ice and snow and therefore makes them easy to find. Most of the meteorites found in the Antarctic have terrestrial ages between 10,000 and 700,000 years, that is, the time they have spent just lying on the ice and snow.

Because meteorites are from objects such as asteroids, comets, moons, and planets, some dating from the very origin of the solar system, some 4.5 billion years ago, scientific interest in them has always been very high. Thus, with their discovery in Antarctica, the U.S. Antarctic Meteorite Program was established as a collaboration between the National Science Foundation, NASA, and the Smithsonian Institution. The National Science Foundation geologists are responsible for collecting specimens, while NASA and the Smithsonian Institute curate them.

Between 1973 and 1996, 18,300 meteorite specimens were gathered by ANSMET (the Antarctic Search for Meteorites) researchers in the Antarctic. With so many of these celestial rocks being discovered, only those that show unique characteristics have started to be extensively analyzed. Meteorites are given designations that indicate where they were found, what year, and what number the specimen was during the year it was found. For example: EETA 79001 translates to a meteorite discovered in the Elephant Moraine, Antarctica region (hence EETA), in the year 1979, and it was the first specimen found that year, 001.

Some of the most intriguing meteorite specimens ever found are the class known as the SNCs, sometimes called "snicks." SNC is an acronym for Shergotty, Nakhla, and Chassigny, representing the places on Earth where the first samples were found. The Shergotty, India, specimen was found on August 25, 1865 (one specimen). The Nakhla, Egypt, specimens were found on June 28, 1911 (more than forty specimens). The Chassigny, France, specimen was discovered on October 3, 1815. For years these meteorite specimens remained an enigma because of their relatively young ages, 200 million to one billion years old, compared to most other meteorites having ages of four billion years. They contain bonded water in some of their minerals.

Today a total of twelve meteorites have been identified that belong to the SNC variety. All have similar characteristics that define them as meteorites from Mars. Depending on slight variations of mineral content, each new SNC meteorite discovered is classified as a Shergotty, Nakhla, or Chassigny. Nine have been found by the U.S. ANSMET Team, and more SNCs are thought to be in the collections of Japanese Antarctic Researchers who seek to study meteors as well. Two of the most important of these are EETA 79001 and ALH 84001, found in Antarctica. It was actually EETA 79001 that got scientists to speculate on a Martian origin for all of the SNC meteorites, because it contained gas bubbles (discovered in 1983 by Drs. Donald D. Bogard and P. Johnson of the Planetary and Earth Sciences Division, NASA Johnson Space Center, Houston, Texas) in its matrix that closely matched the atmospheric composition of Mars as sampled by the *Viking* spacecraft in 1976. In fact, the most unusual aspect of EETA 79001 aside from the gas bubbles matching Mars' atmospheric content is its age—only 175 million years old. In terms of meteorite age, it is just a baby. ALH 84001 was given an age of 4.5 billion years by comparison. It was EETA 79001 that became the Martian SNC meteorite "Rosetta Stone," as it linked the other eleven SNC meteorites to Mars. Although the other eleven SNC meteorites do not contain similar gas bubbles from Mars, they all have the same distinctive oxygen isotope ratio (oxygen-17 to oxygen-18) and elemental makeup with iron-rich silicates and iron-oxides, and these are so closely matched to EETA 79001 that all are thought to originate from the same source—Mars.

In 1984, a National Science Foundation geologist working on the U.S. Antarctic Meteorite Program picked up an unusual greenish-looking meteorite off the ice field near the Allan Hills region of

Antarctica. Dr. Roberta Score found the first Antarctic meteorite specimen of 1984, weighing 4.2 pounds—ALH 84001—while four other fellow researchers were taking a break and having some fun riding on their snowmobiles. Per usual procedure Dr. Score labeled and placed the specimen in a sterile teflon collection bag, along with a hundred other samples gathered during the expedition. These were later sent to be curated at the Johnson Space Center (JSC) in Houston, Texas. There they sat for twelve years until Dr. David McKay, a JSC planetary scientist, and others began to analyze the peculiar specimen. Besides Dr. David McKay, other planetary scientists had requested small pieces of ALH 84001 to analyze.

The first unusual thing discovered about ALH 84001 was its age, thought to be about 4.1 to 4.5 billion years old. This did not agree with any of the young ages of the other SNC meteorites, so it was classified at first as a diogenite meteorite, a type of achondrite meteorite thought to come from the asteroid Vesta. Dr. David W. Mittlefehldt, a planetary scientist working out of the University of Arizona, finally concluded that ALH 84001 could not be from the asteroid Vesta because of elemental ratios between iron to manganese that are characteristic ratios for planets like Earth or Mars.

ALH 84001, world-famous SNC meteorite thought to contain fossilized bacteria from Mars.

Dr. Mittlefehldt sent a chunk of ALH 84001 to Dr. Robert Clayton, a geochemist from the University of Chicago who specialized in conducting oxygen isotope mapping of solar system objects. Clayton's analysis showed that ALH 84001 did not share the same oxygen isotope ratio as the Earth-Moon system or asteroids. After further extensive analysis, Clayton concluded in 1994 that ALH 84001's oxygen isotope ratio fit right in with the rest of the SNC meteorites with one exception—its age, over four billion years old.

Was ALH 84001 with its extreme age really from Mars? After all, 4.5 billion years ago was a time when the planets were thought to be still molten and accreting. Also, many scientists thought the gases in any Mars atmosphere four billion years ago would have to be much different from the gases present in Mars' atmosphere 200 million to three billion years ago, as suggested by most of the other SNC meteorites. This was an astounding discovery all by itself! That Mars' atmosphere would remain unchanged for at least 200 million years (as measured by *Viking*) would have a tremendous influence on the Mars-life issue, as we shall read. Dr. Timothy Swindle of the Lunar and Planetary Institute in Arizona studied ALH 84001 in 1994 and wrote a scientific paper on it stating:

> It is much more difficult to explain ALH 84001 if it is not from Mars. It is telling us things about Mars that we have not figured out yet. It has large chunks of carbonate minerals in it, and the arguments are that they are caused either by weathering, magmatic activity, and most recently—biology.

Dr. David McKay of the JSC, during his preliminary investigation of ALH 84001, found that this unusual SNC specimen seemed to have two separate ages—one for the crystallization age of the lava in it (4.5 billion years), and one for the carbonate inclusions (3.6 billion years). Dr. Meenakshi Wadhwa, a geochemist from the Field Museum in Chicago, disagrees with McKay's age estimate for the carbonates in ALH 84001. After making her own calculations based on ages determined from concentrations of rubidium and strontium in ALH 84001, she stated that the age of the carbonates is much younger, around 1.3 billion years old. This paradox spurred McKay to look at ALH 84001 with extreme care and attention. McKay put together a research team consisting of Dr. Everett Gibson, a JSC planetary scientist; Dr. Kathie Thomas-Keprta of Lockheed-Martin; and a Professor of Chemistry at Stanford University, Dr. Richard Zare. Together, this team of scientists would unravel one of the most spectacular finds in the history of science—evidence of fossil bacteria (inside a meteorite) from another world, Mars.

McKay and his team started work on ALH 84001 in 1994, and by early June of 1996 they had pieced together one of the most interesting puzzles of all time, that life may have existed on Mars 3.6 billion years ago. The chronological history of ALH 84001 is interesting. It is thought to have been blasted off the surface of Mars about 16 million years ago by a large oblique meteorite or asteroid impact, and then spent perhaps millions of years in space before eventually getting pulled into Earth's gravitational field and landing on the ice field of Antarctica 13,000 years ago.

One of the first observations made by McKay's team was that ALH 84001 had easily detectable amounts of organic molecules called polycyclic aromatic hydrocarbons (PAHs), which look like ring structures with six sides each and are made of carbon and hydrogen. ALH 84001 demonstrated evidence of organic material on Mars—something the Viking GCMS could not—but it was not considered evidence for life because PAHs had been found in a host of other meteorites and interplanetary dust particles. Life on Earth is more complicated than that because it contains carbon, nitrogen, hydrogen, oxygen, sulfur, and phosphorus. ALH 84001 only had two of those, carbon and hydrogen. PAHs have been traditionally associated with heat-degraded organic material in meteorites, so PAHs by themselves do not mean life. However, located and concentrated within the vicinity of the PAHs were carbonates. This suggested that Mars had a similar process to the Earth's by which atmospheric carbon dioxide is absorbed by oceans or lakes and gets deposited in sedimentary rocks.

McKay's team of investigators applied the very latest state-of-the-art scientific instruments to analyze ALH 84001, including a high-resolution scanning electron microscope. The team was prompted to use this powerful tool as a result of the recent discovery on Earth of nanobacteria, bacteria several orders of magnitude smaller than any bacteria previously discovered. Another powerful scientific instrument employed to study this Martian meteorite was Stanford University's dual laser mass spectrometer, the most sensitive instrument in the world to look for organic material. By comparison, the dual laser mass spectrometer is millions of times more sensitive than the Viking GCMS sent to Mars over twenty years ago.

ALH 84001 had other interesting objects to reveal. It contained magnetite, which was thought to be on Mars in abundance according to the analysis returned by the *Viking Lander's* Inorganic Analysis experiment (XRFS). Magnetite on Earth can be produced two different ways: by inorganic chemistry involving oxidation of iron, or by the still-not-understood mechanism of biomineralization, a process by which bacteria produce their own magnetite through an enzymatic reaction. Sometimes the Earth bacteria have magnetite on the inside of their cells, and some bacteria have it bound to the outside of their cell walls. Magnetite, like reduced iron, would be an excellent shield against ultraviolet light. Are the bacteria that make magnetite doing so because the ancient Earth, like Mars today, had no ozone shield to prevent powerful UV rays from penetrating their cell walls? On a windy planet like Mars it would be beneficial to organisms blown around and up into the lower atmosphere (just like they do on Earth) to have a UV shield such as magnetite or iron attached to the outside of their cell walls. The difference between inorganic magnetite and biologically produced magnetite is in the shape of its structure. Biogenic magnetite can be hexagonal, cubic, or tear-shaped. The shapes of the magnetite contained in ALH 84001 are the same as those produced on Earth by bacteria in low-temperature fluid environments.

Another exciting mineral found in ALH 84001 was greigite, an iron-sulfide commonly produced on Earth by bacteria. The crystals of greigite in ALH 84001 were four times smaller than their terrestrial counterparts. Since Mars has one-quarter the mass of Earth, evolution of smaller organisms might seem logical. Pyrrhotite, a substance composed of iron and sulfur that can either be produced inorganically or by microorganisms, was also present. The simplest

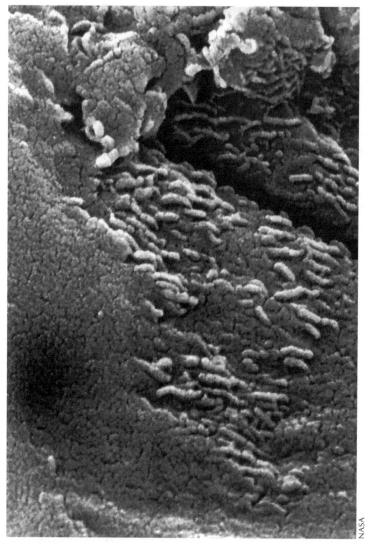

Electron microscope image inside of ALH 84001 with alleged colonies of fossilized bacteria.

explanation for pyrrhotite particles distributed around the rim of the carbonate crystals in ALH 84001 was microorganisms from Mars.

The evidence was getting more interesting by the day. The carbonate (calcium carbonate) globules were covered by magnetite and iron-sulfides mixed in with the PAHs. However, the most startling

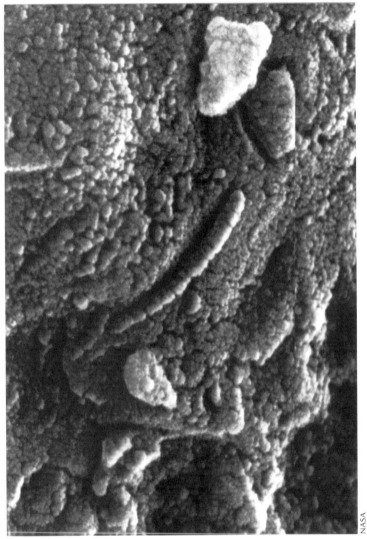

NASA

ALH 84001. Fossil remains of a segmented organism or inorganic debris?

discovery in ALH 84001 were numerous elongated tube-like structures believed to be fossilized bacteria from Mars. McKay and his research team suggested that the carbonate inclusions in ALH 84001 were the products of biologic activity. McKay and his team found the textures of the carbonate globules very similar to "carbonate crystal bundle precipitates" found in freshwater ponds as bacteria alter their environment (McKay, 1996). However, "the fossil-like structures are 100 times smaller than any fossil bacteria found on the Earth," says Dr. William Schopf, Professor of Paleobiology at the University of California. Schopf remains highly skeptical of ALH 84001 having evidence for fossil bacteria from Mars.

Dr. Robert Folk, a geologist from the University of Texas, has claimed to have found two-billion-year-old fossilized bacteria on carbonate rocks from Italy. Folk said the Martian microfossils and Italian microfossils look identical. The environment where Folk found these "nanobacteria" was a hot spring.

Of course, "the rest is history," as the popular saying goes. NASA made the official announcement of evidence for possible life on Mars on a world-wide television broadcast on August 7, 1996. President Clinton himself addressed the issue as being one of the most important discoveries in scientific history. But no one was saying a word about the possibility of current life on Mars.

For years most planetary geologists studying Mars thought that for at least the first billion years of its history, water covered parts of the planet's surface as evidenced in *Mariner 9* and *Viking Orbiter* images. Even though some of this water may have covered large areas of Mars 500 to 1000 meters in depth, their theories today say that because Mars then underwent a catastrophic loss of its atmosphere, it became colder, with a subsequent loss of liquid water as surface water became ice. Thus any life on Mars that may have begun could not evolve with the changing conditions and necessarily became extinct.

The record of our planet's biological history does not share that same theory. Earth had a carbon dioxide atmosphere produced by volcanic activity and microorganisms with no ozone layer to protect life from large doses of ultraviolet radiation, yet life prevailed and flourished. When photosynthetic organisms began pumping oxygen into the Earth's atmosphere it was the single largest environmental disaster in the history of our planet. It must have killed most of the original descendants of microbes that were here initially. Or did that life retreat into the interior of the Earth? Microorganisms are continually being discovered at ever deeper regions—more than a mile.

Either life retreated to the interior of the planet or perhaps was there all along.

Today the oxygen content of our atmosphere combined with other oxidants is thousands of times more oxidizing than is Mars. So why have scientists been saying that life could not evolve to survive on Mars today? Let us examine the most frequently cited reasons for Mars not being able to support life, compare them to terrestrial analogies, and then allow the reader to draw his or her own conclusions.

Water

Probably the single most frequently voiced argument against extant life on Mars is the theory that Mars does not have enough liquid water to sustain an ecosystem. Is that true? I have already covered elsewhere in this book the fact that Mars can and probably does have some liquid water in the first few centimeters of its soil for at least brief periods of up to several hours per day and weeks on end when the barometric pressure is above 6.1 millibars and the temperature above freezing. This is accomplished through a process known as capillary action, when a fluid such as water is drawn up in small interstices or tubes as a result of surface tension. Studies of different soil mechanics on Earth show that fine soils, such as silty clay, have the highest rate of capillary rise. Translated, all this means is that water can travel from a depth of ten meters to the surface in a twenty-four-hour period in silty soils.

The Viking Physical Properties experiment demonstrated that the soil of Mars is similar in grain size to clay silt and showed evidence of what scientists called duricrust (extensively discussed earlier). If you ask a geologist the definition of duricrust, he or she will tell you that thin films of moisture transport dissolved salts such as chlorine, bromine, and sodium chloride through capillary action to the surface of a soil. As the water approaches the surface and begins to evaporate, the salts collect in with soil particles.

If you ask geobiologists their interpretation of the formation of duricrust, they will tell you that it involves the process of biomineralization caused by microorganisms. As mentioned in an earlier chapter, the surface of soil and rocks on Mars can reach temperatures in excess of 90 degrees Fahrenheit (310 degrees Kelvin, personal communication, Dr. Terry Z. Martin) during the summer seasons in equatorial regions. A principle known as thermal osmosis dictates that water at a higher temperature than its surroundings will migrate in the soil towards a cooler area. When the Viking Biology Team sampled soil from under rocks, they indeed found more water, showing that the thermal osmosis principle is active on Mars. The fact that thermal osmosis is a factor on Mars proves that water is variable and moving through the soil.

Where does the water come from? Most likely a number of different sources including the following considerations: Before the Viking Mission to Mars in 1976, scientists assumed the northern and southern polar ice caps consisted mainly of frozen carbon dioxide. After Viking conducted water vapor mapping, it was discovered that the northern polar ice cap was frozen water, since temperatures were above the freezing point of carbon dioxide nearly all Martian year long. The southern polar ice cap also contained a reservoir of water, but it lay underneath a thin sheet of carbon dioxide ice. Temperatures in the southern polar hemisphere of Mars are colder because of the planet's eccentric orbit and axial tilt.

Drs. A. L. Sprague and D. M. Hunten conducted Earth-based studies of Mars from 1991 to 1995 using a 1.5-meter telescope at the Catalina Station on Mount Bigelow of the University of Arizona. There, the astronomers used an echelle spectrograph coupled to the Catalina telescope, incorporating a new CCD detector to observe Mars' atmosphere in the infrared. This new equipment provided exquisite sensitivity, and their data showed a significant departure from the Viking MAWD water vapor measurements made from 1976 to 1979. The Viking MAWD (Mars Atmospheric Water Detector) maximum water vapor measurement found a large 97 perceptible microns in the northern hemisphere, the largest ever recorded. Based on their findings, variation of water vapor abundance within a Martian season is sometimes as large as a factor of 3 from one year to the next, reaching as high as 36.4 perceptible microns in the northern hemisphere, much higher than the MAWD averages. Late spring and summer are the wet seasons for both northern and southern hemispheres, with the northern latitudes being five times wetter than the southern. Perhaps the most striking observation made by these astronomers was that Mars' atmosphere had a water vapor variability on timescales of hours. What could account for such a variation of water in the Martian atmosphere? On Earth, that kind of variability is caused by rain and snow! Were Sprague and his colleagues suggesting rain on Mars? Here is what their paper for the *Journal of Geophysical Research* says:

> We believe a plausible explanation for some of the rapid variability in our measurements is that water vapor is removed from the air and present in clouds as water droplets.... (Sprague, A. L., 1996)

Would the combined water vapor of the northern and southern polar ice caps be enough to support a microbial ecology? Possibly,

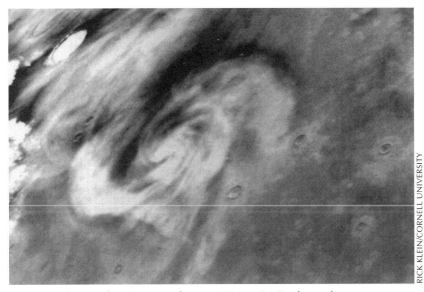

RICK KLEIN/CORNELL UNIVERSITY

Mariner 9 *image showing a cyclone on Mars. On Earth, cyclones occur as a result of a saturated atmosphere over large bodies of water such as the Pacific Ocean. Such storm systems on Mars appear localized to the northern polar region.*

but there is evidence of a greater source of water than just the polar ice caps—it is in the regolith or soil. Many *Viking Orbiter* views of Mars show splash-craters that formed when a meteorite slammed into the surface at a superheated temperature, melted the ice-regolith, and left a giant splash mark instead of a common ejecta blanket. These splash-craters suggest the crust of Mars has substantial amounts of frozen water.

I have mentioned previously that outgassing evidence from the subsurface of Mars is found in some *Mariner 9* and *Viking Orbiter* images. What were thought to be wind-streaks in some northern polar region craters in *Viking Orbiter* photographs have now recently been revealed to be frost-streaks, particularly around dune fields in and around the craters. Obviously, water is freezing out of the thin atmosphere to form these frost patterns. Could a similar process be happening on a global scale? Thousands of Martian craters with "wind-streaks" are found in the lower latitudes. Their albedoes are so bright against the surrounding region, is it possible they might be frost-streaks as well? Evidence exists that Mars may still have a geo-thermal source. Maybe as water is evaporated from the subsurface floors of certain craters, it freezes in the thin, cold, Martian wind and is responsible for the highly elongated, extremely bright streaks.

In 1975 an airborne radio-echo survey was conducted on ice depths in East Antarctica which led to the discovery of a large sub-surface lake almost four kilometers beneath the surface of Antarctic ice. The lake is heated from below by geothermal energy rising to the surface and kept fluid under a thick ice cover. Lake Vostok was found by the satellite ERS-1, designed specifically for studies of large polar ice sheets in 1993. No one in 1975 could imagine just how large the lake was. It was being compared to the Great Lake, Ontario. Mea-surements by the ERS-1 satellite seemed to indicate the lake had a depth of 500 meters. Estimates of the lake's age were placed at 50,000 years. Imagine an underground lake that had not been exposed to the air for more than 50,000 years. What kind of ecosystem awaits eager scientists? Concern over the contamination of Lake Vostok is so great that all precautions are being taken not to contaminate the lake through the use of deep ice core drilling.

What does Lake Vostok have to do with water on Mars? An extensive Earth-based radar survey of Mars using the Jet Propulsion Laboratory's Goldstone Radar Facility from 1971 to 1973 revealed regions on Mars considered to be large areas of liquid water (Zisk, S. H., *Nature*, Vol. 288, 18/25 December 1980) within 50 to 100 centimeters of the surface. One large region was found in Solis Lacus and two smaller areas northwest of the great Hellas Basin. Another area found to have characteristics of subsurface liquid water is Noachis-Hellespontus. With strong evidence of recent geothermal and volcanic activity documented by Dr. Leonard Martin and Dr. B. Lucchitta (mentioned in Chapter Six), the existence of many lakes and rivers under the ice-regolith of Mars remains a possibility.

What kind of life do you find in a dark, cool, underground lake? For that answer, you can ask Dr. Todd O. Stevens and Dr. James P. McKinley, both geomicrobiologists at Pacific Northwest Laboratory in Richland, Washington. Stevens and McKinley, while working (in part) for the Department of Energy's Subsurface Science Program, began testing the groundwater in deep Columbia River basalt aquifers in the state of Washington. The team sampled water from these dark depths up to 1500 meters deep. What they discovered was simply amazing—anaerobic bacteria living off a reaction of nothing more than basalt rock and oxygen-free water. No photosynthesis was required. Just think about the meaning of this discovery: Bacteria with this capability might thrive anywhere in the solar system where liquid water and volcanic rock are available. That is a lot of possibilities! Mars, Jupiter's moons, comets, Saturn's rings and moons, maybe even the Moon or Mercury at certain depths or in deep craters—all could have this type of rock-eating bacteria. *(See Color Plate 22.)*

Drs. Stevens and McKinley named the microbes SLiMEs, an acronym for Subsurface Lithoautotrophic Microbial Ecosystems. There were a number of different species in this dark kingdom, including methanogenic bacteria which produce methane from hydrogen and carbon dioxide; there were sulfate-reducing bacteria that produced hydrogen sulfide from hydrogen and sulfate; acetotgenic bacteria producing acetate (vinegar) from hydrogen and carbon dioxide; and fermentative bacteria that live off the by-products of other bacteria! It was a whole new world unto itself. The bacterial species have yet to be named, as it takes much biochemical and genetic work to identify them. The SLiMEs are among the hot topics with today's microbiologists. SLiMEs are officially designated as chemolithoautotrophic organisms or "rock eaters." Could SLiMEs have been detected in the Viking biology instruments? Dr. Todd Stevens replies, "The Viking life-detection experiments would not have responded to chemolithoautotrophic bacteria. Those experiments were designed to detect photosynthetic bacteria, heterotrophic bacteria, and inherent soil metabolism."

Dr. William Fyfe, an environmental geochemist at the University of Western Ontario in London, is not surprised by the finding of microbes deep within the Earth: "The aggregate bulk of microbes throughout the Earth's crust over the billions of years of geologic time and accumulating, is now viewed as exceeding the mass of the Earth as a whole. That means six sextillion, 588 quintillion tons of them"! Dr. Fyfe's recent work includes the role microorganisms have played in the formation of the Serra Pelada gold field in the Amazon jungle in Brazil. Apparently Fyfe has found a connection involving biomineralization of microbes and the concentrating of gold from jungle soils and rocks! Another microbiologist from the University of Toronto, Dr. Grant Ferris, states that millions of microbes have no problem over geologic time forming silver, gold, carbonates, oxides, sulphides, phosphates, and silicates. One of the hot new areas of scientific research is the study of biomineralization, or how microbes manage to gather inorganic substances inside and outside of their cell walls through enzymatic processes.

Water on Mars exists in the form of subsurface ice, permafrost, and occasionaly, surface liquid water. Ice-fogs are common and by themselves could sustain a surface microbial ecosystem such as the cryptoendolithic algae that bore into rock pores and then use meltwater from ice-fog crystals settling to the surface of substantially heated Martian rocks. As ice-fog forms in the Martian dawn, the Sun can heat the rocks up to temperatures above 90 degrees Fahrenheit, melting ice crystals into the pores of the rock. Fog communities exist on Earth in the Peruvian and Atacama deserts and represent one of the most fragile ecosystems on our world (Rundel, 1991).

The fog-filled craters of Mars. If the fog is just ice-fog crystals as scientists have proposed, why don't they leave behind frost? Fog could be a life-sustaining source of water for a microbial ecosystem on Mars.

Since we know that Mars has water with evidence of geothermal activity and much more sulfur than Earth, we can use our imagination to visualize a vast network of underground sulfur springs containing microorganisms similar to red sulfur bacteria that grow in a strictly mineral media with carbon dioxide as their only source of carbon. These organisms on Earth have carotenoid pigments to gather and utilize light for energy. Martian organisms might have developed a faculative mechanism for living in the dark and in the light. Perhaps on Mars organisms could live chemolithoautotrophically in the dark. When exposed to light, photosynthetic pigments would take over to provide the energy needed for metabolism. Of course, if Martian organisms are as creative as terrestrial organisms, then perhaps they could bring the mountain to Mohammed. How so? Deep-water sponges in the Ross Sea off the coast of Antarctica have spicules that stick out of these animals a distance of more than 100 feet. The spicules act as a natural fiber optic network, directing light down into the dark icy waters and finally down inside the sponge,

271

where colonies of photosynthetic algae live inside its body! The ultimate sponge/symbiont interaction. Nature always seems to find a way around a difficult problem. There could be microbes deep inside Martian sulfur springs that use chemical energy to provide light for photosynthesis. This unique characteristic, well known to many terrestrial organisms, is called bioluminescence but has never been shown to supply light for photosynthesis on Earth.

There are organisms on Earth that can live inside salt crystals for 200 million years and longer, similar to the way bacteria are preserved in Siberian or Antarctic permafrost, slowly metabolizing over antiquity. These extreme halophiles (salt-loving bacteria) were discovered in salt deposits at the Department of Energy's Waste Isolation Pilot Project within their high-intensity storage area in New Mexico. Since then it has been postulated that extreme halophiles would be excellent candidates for current life on Mars, and this is why extreme caution must be exercised in bringing back a sample of soil from Mars. Researchers have found that atmospheric carbon dioxide penetrates down into the crust of salt-lake beds and provides these "endoevaporite" communities with a source to produce their own organic material. Thus even in a seemingly desiccated salt crystal, nature has found a way to keep organisms alive.

Both chlorine and bromine salts have been identified in the Martian soil (as covered earlier). Caliche or "duricrust" was evident at both *Viking* landing sites. Algal strata covering hundreds of thousands of square kilometers in the state of Oklahoma was first discovered by an environmental scientist in 1941, Dr. W. E. Booth. The algal strata were made up of a soil-colored strata, a dark-colored strata, and a reddish-colored strata, all of which contained bacteria and, most importantly, different species of algae (Booth, W. E., 1941). Another desert crust made by biochemical and geochemical processes is the Nari-Lime-Crust, found first in the deserts of Israel. It is a hard calcareous crust that coats different sediments and rocks. Algae and other microbes in these communities live in a very close symbiotic relationship (Krumbein, W. E., 1968). Large colonies of algae live in combination with high numbers of isolated fungi which build up endolithic lichens. Nitrogen required by heterotrophic bacteria would have to be fixed by the green and blue-green algae. On their destruction, the ammonifying bacteria take up the amino acids and reduce them to ammonia. The ammonia will partially diffuse and be partially oxidized by autotrophs. The blue-green algae then become the next generation of algae forming layers in the crust.

The most interesting property of these desert crusts or "caliche" is their ability to switch almost instantly from a dry dormant state to an active state upon wetting. Could such a vast living desert crust lie under the surface of the soil on Mars and be responsible for a planet-wide wave of darkening when atmospheric water vapor or water become available? *(See Color Plate 23.)* So far, the wind-blown dust theory is considered by most scientists to be the cause of the wave of darkening. New spacecraft going to Mars during the next decade should resolve this issue.

It has been known since 1963 by geomicrobiologists that desert varnish, the shiny substance coating many desert rocks, is created through a process involving carbonate solution and the generation of chelating substances produced by algae, lichen, and bacteria (Krumbein, W. E.). As mentioned in an earlier chapter, the Principal Investigator of the Viking Physical Properties Team, Dr. Henry J. Moore, said that many of the Martian rocks seemed to have a coating of desert varnish on them.

Radiation

What would prevent a microbial ecosystem from flourishing once the water problem is solved? What could destroy bacterial spores from the interior of Mars propelled to the surface by out-gassing events? We have already examined the ability of lyophilized bacteria to survive in the vacuum of space and on the Moon (recall *Surveyor III*). Dr. Christopher McKay says he believes radiation would be the limiting factor for life on Mars:

> Not cosmic radiation or solar UV, because you can always imagine burying the organisms a meter or so underground where they

are shielded from cosmic radiation or solar UV. In fact, I think the radiation that would "do them in" is the intrinsic radiation of Martian soil, just like on Earth everywhere. Parts-per-million amounts of uranium, thorium, and potassium in rock and soil provide background radiation. How long does it take to accumulate a lethal dose? For what we know of soil organisms, a lethal dose is about 3,000,000 rads and takes about 10 million years. So, in that case, the limit of survival of a spore sitting on the surface of Mars or Earth for that matter—because there is not that much difference—is the background radiation from trace radionucleotides, and that limit is probably no more than 100 million years.

Dr. Gilbert V. Levin disagrees, citing that McKay has worked with organisms that have been in permafrost for many millions of years, and yet the organisms are viable when thawed. The same principle should apply for intrinsic radiation on Mars. The idea of finding spores still alive on the surface of Mars "doesn't seem possible," McKay added, "unless there is a source—an active vent somewhere under the subsurface that is spewing them out on time scales shorter than millions of years."

One of the most amazing microorganisms on the surface of the Earth has to be the radiation-resistant species known as *Deinococcus radiodurans*, which were discovered living inside the core of the Three Mile Island nuclear reactor in Pennsylvania in 1989. How do these vegetative cells survive? In the process of suffering megarads of radiation and damage while they are growing, certain enzymes produced by the cell are excreted that repair their DNA. So, in a sense what they do is constant damage control. They can only do this when actively metabolizing, not in a dormant, frozen, or desiccated state. A Martian organism with this kind of ability combined with a pigment to protect it from solar UV damage could exist on the surface of the Martian soil in a constant state of repair from intrinsic radiation; it could go through normal changes of dormancy and activity undamaged.

Organisms acclimated to dry and cold environments are more resistant to radiation damage. In addition, it was discovered in 1994 that radiation damage to cells seems to be enhanced in the presence of oxygen. Since Mars has a 95 percent carbon dioxide atmosphere and only a trace amount of oxygen, radiation damage accelerated by oxygen exposure might be reduced.

Carbon Dioxide Atmosphere

As we read earlier in this book, over 90 percent of all the carbon dioxide in Earth's atmosphere is produced by bacteria. Even though the atmosphere of Mars is thin by terrestrial comparison, Mars has a greater amount of carbon dioxide than Earth. The Viking orbital infrared measurements showed that the northern polar ice cap of Mars is too warm to contain much carbon dioxide, and there are

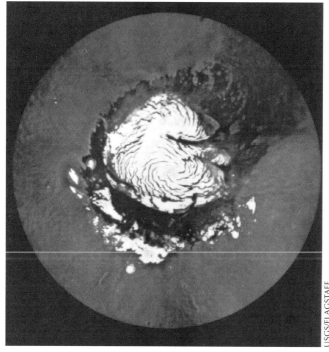

USGS/FLAGSTAFF

The north polar ice cap of Mars was discovered by Viking to contain only water ice, not frozen carbon dioxide as previously thought.

equally compelling reasons to believe that the southern polar ice cap contains too little carbon dioxide to count as a reservoir of atmospheric carbon dioxide.

Where is it coming from? Some scientists have proposed that carbon dioxide is coming out of the Martian soil as a result of being absorbed on soil particles over geologic time. The other possibility that has rarely been discussed is that the carbon dioxide content of the Martian atmosphere is replenished by an active microbial ecosystem.

Dr. E. Imre Friedmann has found that limestone-boring endolithic fungi and lichens occur in terrestrial carbonate rocks using organic acids such as acetate to break down and biologically weather rocks in Antarctica and desert regions. Furthermore, he has recently studied a lime-boring cyanobacterium from Negev Desert rocks called *Matteia sp.* that dissolves carbonate rocks to release carbon dioxide! Friedmann suggested that this type of organism might be used to terraform Mars one day by increasing its concentration of carbon dioxide to the point where sufficient greenhouse warming could take place.

Unfortunately, the microbial dissolution of carbonate rocks, though an active process, is a minor factor in the bacterial carbon dioxide production in Earth's atmosphere. But what if these carbonate-eating types of microbes or others with even greater capabilities exist on Mars today and are responsible for releasing large amounts of carbon dioxide? This process along with other bacteria, fungi, and algae could account for Mars' atmosphere. What about Martian evidence of carbonates? According to the results published by Dr. David McKay and his Research Team in the September 16, 1996, issue of *Science* on the ALH 84001 Martian meteorite, indications are that carbonates might have been plentiful on Mars, generated in part by colonies of once-living organisms that existed in a fluid environment.

Evidence for a Martian Microbial Ecosystem

Taken along with Dr. Gilbert V. Levin's and Dr. Patricia Ann Straat's nine LR experiments run on Mars in 1976, and the new photographs of Mars coming out of the NASA Jet Propulsion Laboratory (JPL) and the Hubble Space Telescope, there is evidence for a planet-wide microbial ecosystem to exist on Mars. The soil of Mars is no longer portrayed as the NASA-tweaked orange-red oxidized-iron landscape that was first advertised during Viking. The soil is in fact more brown than orange or red. The rocks have numerous colors ranging from gray to blue-black with olive, yellow, and brown spots on them that seem to intensify with the seasons. These spots seem to extend a few millimeters above the rock surface they are attached to and reappear even when covered by a few microns of dust settling from the atmosphere. What are they? Something similar to lichens, as Gil Levin suggested in 1976?

Nor is the sky of Mars simply a pinkish-salmon for long periods of time. It is instead known to be variable as dust settles out of the atmosphere. The dust particles that in Earth's atmosphere give us red sunsets produce similar effects on Mars, except for occasions when planetary dust storms occur and linger. Then, as I mentioned in Chapter Five regarding my conversation with Dr. Philip James, the sky color gradually changes as the dust settles completely out of the atmosphere, shifting from pink to yellow to light blue to dark blue and finally to purple, as will be witnessed (ideally) when NASA's *Mars Pathfinder* turns its color Lander camera skyward. Also, NASA's *Global Surveyor Orbiter* will have a sophisticated color camera capable of resolving objects on the planet as small as 1.5 meters. It is important to remember when trying to understand the colors of Mars that the planet's atmosphere can be dust-free for years on end.

To understand a Martian ecosystem and its possible development, we should look at how an ecosystem is built on the Earth. The process, known to environmental ecologists as succession, begins with bare rock. Lichen, "the pioneers of vegetation," establish themselves on the rock. In time, other air-borne organisms cover the rock. They begin to break down the surface of the rock into tiny particles using organic acids they secrete through the process known as biological weathering. This might be the point on Mars where succession has halted and remained because of limited amounts of water (compared to Earth) and colder temperatures. On Earth, succession would continue as soil begins to form and mosses take root. When the soil depth becomes great enough, grasses begin to grow. Their decomposition further adds to the nutrient base in the soil. This allows for shrubs and trees to be introduced. The trees slowly become dominant over the shrubs. As this develops into a mixed mature forest, it supports many different kinds of trees. As the trees age, however, one species becomes dominant and what is left is a mature mixed forest and climax community. The vegetation that remains in the climax community is largely dependent on environmental factors including soil nutrients, liquid and atmospheric water vapor, and annual temperature range.

From this Earth analogy we could conclude that any Martian microbial ecosystem over time might have produced the depth of soil that was dug up by the *Viking Lander* Sampler Arm—at least 14 centimeters and it probably goes much deeper. It is true that other processes such as volcanic activity and wind erosion can and do weather rocks over geologic time; however, the overall majority of all Earth soils are mainly the results of a breakdown by biological processes.

PHOTO BY THE AUTHOR

Algae and lichens on an Earth rock. This is the first step to making soil, a process ecologists have termed "succession."

RICK KLEIN/CORNELL UNIVERSITY

Viking Lander *trench marks where sampler arm dug for soil. Was the soil on Mars a result of biological weathering or just wind erosion?*

Because of the lack of sand particles found by the *Viking Landers* at landing sites 4,000 miles apart, it might be very likely that the greater mechanism for weathering rocks and soil is biological activity. Mars organisms may be very efficient rock eaters, as evidenced by the numerous holes or pits in the rocks at the *Viking* landing sites, generally believed to be volcanic vesicles.

But as I reported earlier in this book, scientists were dismayed by how many holes and pits these rocks and boulders had. Did they resemble anything on the Earth? While these etch-pits may be initially started by melt-water from water-frost or snow deposited on rocks and then constant pelting of ice-fog crystals over geologic time periods, organic acids produced by bacteria and other microbes might be another reason why Martian rocks have more pits and holes than do their Earthly counterparts. In short, Martian microbes may secrete much more concentrated organic acids to bore into the rock. On Earth, many endolithic organisms leave boreholes as they "eat" their way into rock, sometimes dissolving so much they create cavities known as chasms. These organisms then become "chasmoliths." The

279

RICK KLEIN/CORNELL UNIVERSITY

Are the numerous "etch-pits" in Martian rocks volcanic in origin? Could certain types of microbes excrete strong organic acids and literally "eat away" softer portions of these rocks? Rock-boring microorganisms on Earth are common.

PHOTO BY THE AUTHOR

Earth boulder with large etch-pits in which mosses live. Over geologic time, these mosses can dissolve the rock, breaking it down into soil particles.

holes they leave in rocks look very much like some of the pitted and fluted rocks of Mars. Of course, to anyone who has ever gone to an aquarium, some Mars rocks also bear a striking resemblance to "living rock," otherwise known as sea coral. Sea coral, when desiccated and displayed in stores selling aquarium supplies, has many shapes, holes, and colors!

The rocks at Utopia Planitia showed the most striking resemblance to mud-stone rocks or sedimentary sandstones into which shallow water-dwelling organisms "burrow" or "bore" looking for food or protection. Where were the sedimentary geologists at NASA? Did the *Viking 2* landing site hold the greatest clue yet for widespread life on Mars? One large process of erosion on Earth that is seemingly minor on Mars is the wind. Because of the thinness of the Martian atmosphere, wind speeds of up to 110 miles per hour would only feel like a slight breeze on Mars. On Earth we see rocks known as ventifacts that have been smoothed and polished due to constant wind and sand blasting over millions of years. However, ventifacts are rare in the Viking photographs, suggesting that wind erosion played a minor role in the breakdown of soil particles. On Mars, could the wind-blown pelting of ice crystals very slowly weather rocks?

Chapter Epilogue

While we wait for the new spacecraft data from *Global Surveyor* and *Mars Pathfinder*, we must reconsider the evidence for extant life on Mars and now admit it is becoming significant. We must send another set of life-detection experiments and proceed into the twenty-first century with a new goal—to study and protect any Martian microbial ecosystem as well as protect our planet from any possible pathogenic bacteria or viruses from Mars. *(See Color Plate 24.)*

A Mars mission to the old *Viking Lander 1* and *2* sites would prove very interesting, because then we could perhaps see what impact any Earth organisms hidden within the *Viking Landers* have had on these pioneering spacecraft and the surrounding Martian environment. True, Viking was sterilized to the best possible capability, but that does not mean that every microorganism was destroyed. The *Surveyor III* lunar probe was also given a complete sterilization and an organism inside remained dormant for two and a half years on the Moon. To know for certain what changes in the environment of Mars have occurred near the *Viking Landers* would be of enormous scientific interest. Of course, it could also be disquieting for some to see Martian life-forms growing on or attached to the Landers. It would be an ambitious scientific mission.

As far as analyzing a Martian soil sample, an International Space Station might be a solution, but the temptation to bring it back down to Earth may be too great from Earth orbit. Our old friend the Moon has been waiting for us to return to her one day—perhaps now is the time. The study of Martian life on a lunar base specifically designed for that purpose or an equally well-designed lunar orbiting space laboratory would a great international goal. Any Martian organisms—while possibly being pathogenic—may also shed light on the secrets of living systems, maybe even leading us into a new age of medicine. We won't know until we sit down and talk about it sensibly. There can be no deception with something as important as life on other worlds. It is a decision only people representing our entire planet can make because it will take long-term commitment and great financial resources while at the same time creating countless employment opportunities in every field of science.

The search for life and its secrets on Mars could provide the greatest push for education that has ever been witnessed in any society. The interest in life on Mars and life in the universe has always fascinated, and we stand on the verge of comprehending a new reality, that life is the rule rather than the exception. As astronomers discover new planets orbiting distant stars, totalling some fifty new planets as of this writing, it is only a matter of a few more years before the technology to examine the atmospheres in these remote planetary systems will be available. Already scientists are able to ascertain the temperature range that exists for some of these worlds by calculating the distance from their suns. Just like those early pioneers such as Percival Lowell, Gerard Kuiper, and William Sinton, who from their Earth-based observatories searched the planet Mars for signs of life, others will continue in that grand tradition. With the right equipment and the right planet, we will begin to accumulate evidence for life. We are on the threshold of a new era.

ALH 84001 has prompted scientists to begin searching inside other SNC meteorites. Evidence is now building for SNC meteorite EETA 79001 to contain organic compounds that are compatible with a biological origin. The team of British scientists involved in this project includes Dr. Ian Wright and Dr. Colin Pillinger of the Planetary Sciences Institute at Open University in Keynes, along with Dr. Monica Grady of the Natural History Museum in London.

What makes EETA 79001 an even more spectacular find than ALH 84001 is its young age, about 175 million years old. Scientists know that EETA 79001 was blasted off the surface of Mars 600,000 years ago. What relevance does this have? Dr. Michael Meyer, Discipline Scientist for NASA's Exobiology Program, now says, "It means we have more than one sample with hints of life. And it means Mars could have been inhabited for quite a long period and could even have some life today." The Mars life that Michael Meyer suggests might be there today was likely discovered by Dr. Gilbert V. Levin and Dr. Patricia Ann Straat with the Viking Labeled Release experiment in 1976. Both ALH 84001 and EETA 79001 contain convincing evidence (though some are trying to prove otherwise—of course!) that Mars once had organic material, and most likely, Mars has it today as well. It also shows that the Viking GCMS could not detect the organic matter comprising and associated with life, for whatever reasons, and that immediate action should be taken by NASA to send another set of life-detection experiments to determine the facts once and for all.

It is of interest to note that in September of 1996, a month after NASA's announcement of possible past life on Mars, they made another announcement barely noticed by the press. For years NASA's focus for Mars in the next century had been human exploration; now suddenly, NASA reversed this decision, saying only robots will explore the Red Planet. Why? With the discovery of past evidence of life on Mars, NASA surely is aware of the potential of living organisms on Mars today, whether imbedded in salt crystals, surviving in permafrost, or living freely in the soil. To send a human crew first, before making certain the Martian soil did not contain pathogenic organisms, would be suicide.

Only time will tell if evidence is found for fossilized bacteria in a meteorite "not" from Mars, "not" from a planet, but a meteorite from the time when the solar system began to form in a ring of debris around the Sun. Evidence of fossilized structures was found as far back as the late nineteenth century, but was regarded as dubious by most scientists. Today, with powerful new scientific tools such as electron microscopes and the dual laser mass spectrometer, other meteorites are going to reveal their secrets.

Why should this be so significant? It could mean that living matter and inanimate matter may have formed independently but at the same time period in the solar nebula. Just think of the possible ramifications of that scenario. Many problems and questions have been raised about the missing dark matter in galaxies—up to 90 percent. How interesting it would be to discover that perhaps dark matter itself is a giant reservoir of ice, rock, and lyophilized organisms. Since Darwin, biologists have been building a "General Theory of Biology" to explain the mystery of life. The order in which this theory develops is: primordial matter, the elements, compounds, life precursors, single-celled living organisms, differentiated organisms, and finally, consciousness and intelligence. The thrust of this bold hypothesis has been that life is a property acquired by matter at a certain stage of its organization. We now stand ready to witness this hypothesis become fact, and a direct consequence of the General Theory of Biology will be the knowledge that "life" will get deposited and then evolve independently on countless planets that lie in the "habitable zone" of their mother stars. We may even find that life—and indeed, the universe—never had an origin, but rather has always "just been."

With our biocentric need for Earth to be the only oasis for life finally gone, we can then explore the living cosmos with an awareness never before experienced. With the realization that the universe is filled with living entities, we must embark on any exploration of it with great care and responsibility, but we must take the first few brave steps of this new journey into our own "backyard" to fully understand what may be lying in the outer reaches of the wilderness of the cosmos. Our sister world, Mars invites us to come and gain knowledge from her, but not at the expense of either her or our Earth. That ultimate responsibility of protecting life on other worlds and seeding those that do not have it now lies with us. Dr. Robert Williams, Director of the Hubble Space Telescope Science Institute, recently said: "We all have the same origins, we are truly kindred spirits. We are one, not just with each other, but with everything that we see. We are all brothers of the boulders and cousins of the clouds. When we see the stars, we see ourselves."

CHAPTER NINE

Life After Viking: The Evidence Mounts

by Gilbert V. Levin
Biospherics Incorporated

B arry DiGregorio first interviewed me by phone in 1993 for an article about life on Mars. I quickly detected his fascination with the subject. Obviously, he wasn't just writing another freelance, earn-a-living piece. When his story was published in *Final Frontier* in June, 1993, his intense interest in what many have declared to be mankind's most intriguing question—are we alone?—became even more evident. The article, in the form of a Q&A interview, was cogently designed by Barry to call attention to the findings of the Labeled Release experiment on Mars, which he felt had been given short shrift by most of the Viking scientists and had not been adequately disclosed to the public. Barry stayed in touch with me over the next three years, telling me in 1996 of his project to write this book. He informed me that my encounter with Mars through the LR experiment aboard NASA's 1976 Viking Mission would play a prominent role in his book. Barry interviewed me frequently, gathering material for the book and requesting various reprints of scientific publications on Mars written by me and my colleagues. When he came to meet me in person, his interest in the Red Planet and its attendant scientific issues constantly shone through.

However, I already had come to appreciate his dedication to Mars and his desire to settle the life issue. About a year before he started this book, Barry asked my advice when he was preparing his proposal to the Hubble Space Science Institute. Barry proposed to view Mars through the Hubble Space Telescope, the world's most powerful, and, using the highly sensitive Goddard High Resolution Spectrograph, look for hydrogen peroxide in the atmosphere and on the surface of the planet. Unfortunately, the scientific rationale for Barry's experiment was not deemed sufficient by the reviewers for its selection. Barry missed his chance to participate in what pre-Viking

NASA had stated could be "the most important experiment in the history of science," the search for life on Mars.

But why look for hydrogen peroxide on Mars? In one of the classic quirks of science, the very data from the LR life-detection test had been seized by other scientists to contend that the LR had detected not life, but chemicals. They concluded that the LR experiment produced "no evidence for life," but that it did establish the presence of chemical oxidants on the surface of Mars. It did not seem to matter that the LR had been approved as a life-detection test by four NASA-appointed science panels and that the experiment never had produced a false positive result from chemicals encountered in the many Earth soils tested. On Mars, putative chemicals seemed easier to invoke than did life.

The chemistry advocates constructed various theories, beginning with the "raining" of hydrogen peroxide from the Martian atmosphere onto the planet's surface. They contended, and still contend, that the hydrogen peroxide destroyed any organic matter, including life. This oxidant, they explained, either remained in the soil or, depending on the specific theory, complexed with one or more minerals in the soil to form strong oxidants that mimicked life in the LR experiment. The hydrogen peroxide theory also was proposed to account for the "absence" of organic matter in the soil of Mars as reported by the Viking Gas Chromatograph Mass Spectrometer (GCMS). However, as I explain later, there is no evidence for hydrogen peroxide on Mars. Moreover, there is evidence for life existing in an Earth environment in which hydrogen peroxide levels exceed those the chemistry advocates predict for Mars! Other theories to attribute the Mars LR result to non-biological reactions suffer from similar fundamental problems.

Dr. Patricia Straat, my Viking LR Co-Experimenter, and I initially stated and subsequently maintained that our LR data "are consistent with a biological answer." After continuing analysis of the data, and years of additional laboratory testing, I strengthened my conclusion. In 1986, in a talk at the National Academy of Sciences celebrating the tenth anniversary of Viking, I said that "more probably than not" the LR had detected microbial life in the soil of Mars. This produced near-pandemonium among the scientific audience. At the reception, which followed the talks, prominent Viking scientists accused me of having disgraced myself and science. Now, as you will be the first to learn, the facts force me to go a step beyond my 1986 statement.

About a year prior to NASA's announcement of the startling new evidence of biological fossils in the SNC meteorites believed to come from Mars, Barry called me to say he was going to write this book. He then paid me a visit to discuss it. During our conversation, he revealed that the LR experiment and I were going to be featured in his book. This placed me in a dilemma. Since I often have been mis-quoted by those anxious to attribute to me more than I have said, I feared that the careful line I have trod since Viking might be com-promised if I paid no attention to the evolving manuscript. Techni-cal errors might inadvertently creep into text written by even the most well-intentioned non-scientist. The other horn of the dilemma was that, if I did review the work, I would be deemed liable for any opinions or implications that some scientists might think beyond the pale. Having spoken with Barry, I knew his enthusiasm would be dif-ficult to diminish, nor should I attempt to interfere with his views as author. After wrestling with the problem, I offered to review the chap-ters with the understanding that my factual corrections about the Viking LR experiment and related matters would be respected, but that I would have no other control over Barry's literary license, includ-ing his evaluation of the LR experiment and its scientists.

I think Barry has done a fine job in bringing together many dis-crete aspects of Mars and synthesizing them into a fascinating story about our neighbor's capacity to harbor life. His research of the lit-erature has brought forth new insights. I find his snow algae discus-sion particularly interesting and, perhaps, very insightful. I have enjoyed reviewing his work and hope that I helped with the accuracy of the explanations of some of the scientific issues. This book is a popular version of the broad sweep of history surrounding the enigma of Mars, which has cloaked itself more with each attempt to penetrate its mysteries. Using my records to aid my memory, I have tried to keep the *Viking* story straight while, at the same time, realizing that Barry is writing for the general reader and is not attempting a scholarly work. Accordingly, while making some suggestions that I thought would allow scientists to enjoy the book rather than complain of scientific errors, I have resisted my impulses to make it conform to a style more suited for peer-reviewed journals. My disclaimer, however, is that I limit my responsibility for strict accuracy to those parts about *Viking* and its experiments relevant to the search for life, and other pertinent scientific findings discussed in this chapter. I hope the book might promote the fair hearing that the Mars LR data has never received.

After reading Barry's account of my long travail, some will wonder why I "persist" in avid pursuit of the life-on-Mars issue after twenty years; why I do not concede the issue and just go away. It is not a matter of my persisting, the data are what persist. An objective scientist, I cannot change my mind for expedience or comfort.

Viking Revisited

In my 1986 talk at the National Academy of Sciences' tenth anniversary of *Viking* I presented all the available evidence bearing on the LR results. I concluded my analysis with the statement that "more likely than not, the LR discovered life on Mars." Following the publication of news about the first Martian meteorite (ALH 84001) evidence, I told a calling reporter that, were the analyses confirmed, I would change my conclusion to "most likely." When the second meteorite (EETA 79001) results were announced, upon inquiry by another reporter, I said (again, presuming the report valid) "almost certainly."

Now, ten years after my 1986 analysis, I think new events and knowledge require another evaluation. In so doing, I will try to address each aspect bearing on the Mars LR data and other relevant data.

I. The LR Results

In all, nine LR tests were conducted on Mars at two landing sites 4,000 miles apart. All nine supported the finding of living microorganisms. Positive responses, similar to those obtained from Earth soils, were obtained at each location. Heating at 160 degrees Celsius for three hours destroyed the agent. When Viking scientists re-thought that control measure and proposed 50 degrees Celsius as the temperature to distinguish between chemistry and biology, two attempts were made from Earth to direct the instrument to achieve that temperature. One achieved 46 degrees Celsius, which produced a 70-percent reduction in the LR response, satisfying the new and more stringent agreed-upon discriminator. The other such test brought the soil to 51 degrees Celsius, resulting in more than a 90-percent reduction in the response. Terrestrial organisms show such narrow temperature discrimination. Fecal coliforms, for example, are distinguished from other strains in the coliform species by surviving incubation at 44 degrees Celsius whereas the other coliforms, normally cultured at 35 degrees Celsius, do not.

Alas, faced with the new results, the chemistry proponents, still constituting the majority, simply reneged on their agreement that a major diminution in response from soil heated to 50 degrees Celsius constituted evidence that the unheated response had been biologic. They remained immutable even when LR tests on previously active soil samples stored two and three months in the dark sample distributor box at 7 to 10 degrees Celsius produced no response! Microorganisms taken from their natural environment, deprived of their diurnal cycles, might have died.

As stated earlier, another non-biological theory was quickly struck down. It claimed that ultraviolet light hitting the Martian surface "activated" it to cause a physical reaction, releasing gas from within the LR compounds. Again, we resorted to directing *Viking* from the Earth. Commands were sent by which, just at dawn, the *Viking* arm stealthily moved a rock and snatched a sample from beneath it before daylight could "activate" it. This sample, protected from light for eons, produced a positive response within the range of the other positives!

There was only one result of the Mars LR experiment that could be interpreted against biology. The experiment called for a second dose of the radioactive solution to be squirted onto active soil samples after the responses from the first doses had plateaued. Were life present, a sharp increase in the evolution of gas was expected. Instead, when the second dose was added, 20 percent of the gas that had already been evolved disappeared from the space above the sample, re-absorbed into the soil below! The gas slowly re-evolved over about a month. Chemistry buffs contended that this showed that the active chemical in the soil had been used up by the first dose and that the freshly moistened soil had re-absorbed the gas. Had a second pulse of gas evolved, they probably would have said that the first dose of LR compounds had been exhausted by the chemicals in the soil and that the second dose merely supplied fresh reagent to restart the reaction. In an LR test we later performed on lichen in our laboratory, we got a positive response but, with excessive wetting, observed a large re-absorption of the gas evolved. Excess moisture is known to kill these symbiotic organisms. The wetting of an alkaline soil (such as the Martian soil was indicated to be by *Viking* analysis) results in absorption of carbon dioxide. We demonstrated this reaction, achieving the same 20 percent reabsorption of gas, in our laboratory LR instrument. Might the same explanation apply to the result on Mars, especially when lichen offer a model for possible Martian life forms?

2. The GEx Results

The GEx experiment had two stages: first, the "chicken soup" nutrient was uncovered and only the water vapor emanating from it was allowed to contact the soil in the test cell; in the second stage, the soil was wetted with the soup to promote metabolism by any organisms present. The water vapor stage produced a large burst of oxygen from the Martian soil. Even though no light shone on the soil during the experiment, the result was held out as possible evidence of microbial photosynthesis. When the soil was wetted with the GEx liquid medium, carbon dioxide was rapidly absorbed from the atmosphere into the soil but no additional oxygen evolved. When a duplicate sample of the soil was heated as a control and then tested, it still produced the large pulse of oxygen upon humidification. The GEx Experimenter and the Biology Team concluded that GEx had produced no indication of life, but of some chemical oxidant in the soil that reacted with water vapor to yield oxygen. The possibility that oxygen adsorbed on the soil (from the sparse amount in the Martian atmosphere) was released by the GEx vapor also was raised, but discounted because of the large amount of oxygen released in GEx.

3. The PR Results

The PR experiment produces and counts two signals, or "peaks." The first is from the radioactive carbon monoxide and carbon dioxide which had been adsorbed onto soil particles in the test cell. These adsorbed gases are released from the soil by modest heating. The second peak comes from gases released on further heating of the soil to combustion temperature. This signal indicates the burning of organic matter formed during the test. If found, the second peak is the evidence for life.

When the PR's turn came, it produced a very small second peak response, so small that, in Earth tests, it would have been discounted as "noise" by Dr. Horowitz and his Co-Experimenters. Yet this was the big test on Mars, and Horowitz wanted to explore every possibility that he might have detected life. He had the PR instrument re-count the faint signal for a full day to demonstrate that it was statistically significant above the background noise. He published this and called the response "startling." However, he cautioned that "a biological interpretation of the results is unlikely in view of the thermostability of the reaction." The heated control gave the same results as the test sample. After suggesting several non-biological reasons for the results, Horowitz, nonetheless, stated that "it remains to be seen whether any of the proposed mechanisms can account for the intriguing observations," thereby leaving the door to life open.

While the PR signal was above the background level, it was far short of being significant in indicating life. Indeed, tests of the PR on Earth, both before and after the test on Mars, showed even higher responses from control tests with sterilized soils, despite the presence of the UV filter that had been installed to prevent false positives. Apparently, the PR's UV filter did not completely fulfill its mission of removing the rays that caused photo-chemical production of organic matter from Martian atmospheric gases. Thus, the PR experiment gave no evidence for life. Nonetheless, the PR was a very important experiment which, until now, has not been given due credit. It confirmed on Mars that—as in the PR tests on Earth—organic matter is formed in the sunlit atmosphere and, even under the continuous shining of the UV light, accumulates in the soil where the Viking GCMS should have found it! Had these gases not been incorporated into organic matter, they would have been blown out of the test cell during the first peak heating cycle.

4. The GCMS Results

The LR group held its breath while the GCMS went through its motions. And, because of mechanical and design problems, it had to perform sample acquisition twice before Klaus Biemann felt the instrument had obtained a sample. As stated, the Viking scientists all felt certain that there were organic compounds on Mars. Thus the mission of the GCMS was not to *detect* them, but to *identify* them. Our LR group hoped that compounds implicating life would be found. When the analysis was finished, the results astonished everyone. No trace of Martian organic compounds was found!

However, the GCMS had problems which raise questions about the validity of its findings. It could not detect some organic matter of biologic origin in soils only sparsely populated with microorganisms. Mechanical problems on Mars resulted in difficulty in obtaining soil samples. The mechanical difficulty was exacerbated by the fact that the instrument had no "tell tale" to signal the receipt of a sample. The only evidence for samples was the observation that the sampling arm had scraped a small ditch in the soil to obtain the sample. However, there was no way to tell whether that sample made its way through the distribution box, which received it, and then into the tiny ovens of the GCMS. And, if it did, the amounts that entered the ovens were uncertain. It finally was decided that the GCMS had obtained samples because of the amounts of water and carbon dioxide that evolved during the heating of the samples. However, carbon dioxide constitutes 95 percent of the Martian atmosphere, and that atmosphere daily reaches 100 percent relative humidity. Since frost was deposited on the surface daily, it also might have been deposited into the oven and sampling train of the GCMS, along with any carbon dioxide that had dissolved in the water vapor. To what extent these possible deposits might have figured in the GCMS results, particularly if no sample of soil had been obtained, is unknown.

Nonetheless, the GCMS' unexpected result forced a dilemma on NASA and its community of scientists: how could any thought of life be entertained in the absence of organic matter? The easy and cautious way out was apparent. The LR had detected inorganic chemical(s), not life. Once set, that stage has never been changed.

5. Simulations of the LR Results

In the twenty years since *Viking* many attempts have been made in various laboratories to duplicate the LR Mars results by nonbiological means. Our own laboratory spent three years in this effort. Hydrogen peroxide, superoxides, metalloperoxides, peroxide complexes, UV light, and ionizing radiation were tested against Mars analog soils prepared by NASA based on *Viking* analyses of Martian soil, various clays, minerals, and other surrogate soil substrates. We applied a wide range of environmental conditions to the test procedure. LR radioactive solution and its single components were applied to the samples in a Viking-type LR instrument. A wide range of control regimens was used. Under extreme conditions unrealistic for Mars we were able to force positive results. However, no simulation of the Mars LR data could be produced in any of our experiments or those of others when materials and conditions known to obtain on Mars were used. We have published on all of our efforts and on those of others that have been published or otherwise come to our attention. A plausible reproduction of the Mars LR data by nonbiological means remains to be demonstrated.

6. Imaging

Frustrated at the lack of progress in gaining acceptance of the LR results, in 1977 I went to JPL and examined all 10,000 *Viking Lander* images then in the JPL files. The JPL Viking Imaging System staff helped me by producing the digitized images in "Radcam," which means true color calibration. The first thing I noticed was that Mars was not "uniformly orange-red" as stated in NASA press releases. Just as was shown in the first color image released, and then promptly withdrawn, by NASA, the landscape appeared very familiar, very Arizona-like. Generally reddish brown to brown, the landscape contained areas of ochre, yellow, and olive. Most surprising to me were olive to yellow-green to greenish colored areas on many of the rocks.

One night I made the discovery that, on some of the rocks, these colored areas appeared to show changes in pattern and coloration from Martian year to year. I thought it was the second time I had discovered life on Mars! I was so excited that I left JPL and drove up Angels Crest Highway to park and study the night sky until my high wore off.

Analysis of the six channels of digital information comprising each Viking image—red, blue, green, and three near-infra red frequencies—showed these spots on the rocks to be the greenest objects in the entire field of view. Thinking that the spots might look like lichen, on one trip I brought some rocks bearing patches of lichen from Maryland to JPL. I placed the rocks in the *Viking 1* simulated landing site. The JPL Viking Imaging System Team took images of the rocks through the Viking Lander Camera Imaging System. The pictures were taken under the simulated Martian light bathing the scene. They were processed in an identical manner to those obtained on Mars. The digital spectroscopic analysis showed that the lichen on the rocks were the greenest objects in view. Furthermore, the digital values of the color, hue, and saturation were very close to those for the greenish spots on the Mars rocks. Publication of this information in a technical paper had the reverse of the action I anticipated. Instead of supporting the LR biological interpretation, the publication was viewed as a desperate, non-scientific ploy. First, it was widely denied that any greenish coloration showed on the photographs! Next, I was chastised for supposedly intimating that there were lichen on Mars by claiming to have found green areas.

While working on the images at JPL, I called their attention to one of the Viking scientists. He subsequently wrote and published a report on the finding of greenish colored soil and markings on the rocks on Mars. Wary of the sinkhole that awaited biological references to Mars, he made no mention, among the many possible origins he cited for the coloration, of any biological possibility. He even was careful not to reference our earlier paper, nor credit my having called his attention to the colored spots and areas. However, it only has been since publication of his report that the presence of green spots on Martian rocks has become accepted. Recent renderings of the Martian landscape, once again, look very much like Arizona—as demonstrated by the carefully prepared image serving as the cover on the book *Mars*, published in 1992.

7. Liquid Water

The perceived lack of liquid water on the surface of Mars has led many scientists to conclude that life could not be sustained there and, therefore, the LR results must be ascribed to chemistry. "Water is life" has become a popular rallying call of the chemical protagonists. I do not believe the answer to be as simple as that didacticism. As with so many other questions put to Mars, it does not give a straightforward answer with respect to liquid water. The "triple point" for water is 6.1 millibars (mb) of atmospheric pressure. If the total atmospheric pressure exceeds 6.1 mb, water may exist in any of its three forms: solid, liquid, or vapor, depending on its temperature. However, if the total atmospheric pressure is below 6.1 mb, water cannot exist as a liquid. Atmospheric pressure on Mars varies between approximately 6 and 10 mb. Thus, at times when 6.1 mb are exceeded, should the temperature rise sufficiently to melt the abundant ice, liquid water would result. Temperatures of the sampling arms in contact with the soil were recorded at both Viking landing sites. The temperatures of the arms rose as the Sun rose to and a little beyond its zenith. The temperatures of the arm at the *Viking 2* site reached 273 degrees Kelvin (0 degrees Celsius, the melting point of water) and stayed there for a while. That means that liquid water was present under the arms. When water transitions from solid to liquid, just before it melts, extra heat (the heat of fusion) is required before the temperature can continue to rise. This pause in the temperature rise is what Viking recorded—proof of liquid water. It is true that the metal sampling arms absorbed and stored more heat from the Sun than the soil would otherwise. However, it is quite likely that dark rocks, and perhaps dark soil, at the sites of both Landers act the same way. They could supply liquid water to microorganisms if only for a brief period daily. Mars microorganisms may well have adapted to garner needed water in this fashion. Furthermore, a new book (*Water On Mars*, M. Carr, 1996) reports that the *Viking Orbiters* found surface temperatures reaching 298 degrees Kelvin at the summer solstice at 1 P.M. local time in the southern hemisphere, with atmospheric pressures at both Viking sites exceeding the triple point for the 700 consecutive Martian days of measurement. Viking also reported temperatures exceeding 273 degrees Kelvin (98 degrees Fahrenheit) in the northern hemisphere where both Vikings landed.

At the other extreme is the possibility that Martian organisms may be able to obtain their water from the atmosphere, as has been reported for some lichen on Earth. While the Martian atmosphere is only about one percent of ours, both Viking sites showed high relative humidities, reaching 100 percent nightly. Even if the water vapor were in the form of tiny ice crystals, these would deposit on the organisms. The organisms might have learned to store energy from the Sun to melt the crystals and absorb the liquid, or, as many Earth microorganisms do to prevent freezing, Martian organisms might make antifreeze. Finally, ice crystals have been shown to participate in chemical reactions, such as in the atmospheric destruction of ozone by chlorofluorohydrocarbons (CFC), which is responsible for our ozone hole. It is believed that one end of an ice crystal in a cloud remains solid while the end in contact with the ozone and CFC behaves like a quasi-liquid and permits the "aqueous" reaction to take place. This process may be used by microorganisms on Mars (and, perhaps, on Earth, too!).

In sum, there is evidence for liquid water on Mars. In addition, Martian organisms may be able to absorb water vapor, or may have evolved to make liquid water from ice or to take advantage of ice's ability to provide an aqueous environment for reactions. Our knowledge of the water issue—on Earth or Mars—is too uncertain to be used as an absolute barrier to life on Mars.

8. The Martian Meteorites

When meteorite ALH 84001, believed to have come from Mars, was found to contain fossils indicating life, the news startled the world. Once again, the question of life on Mars became front-page news. Analysis indicated the meteorite to be approximately 3.5 billion years old, formed about one billion years after the planet Mars coalesced. By then the planet had had time to cool and become environmentally conducive to life. Liquid water is believed to have been abundant on Mars at that time. This history closely paralleled that of Earth, which is believed to have given rise to living organisms within the first billion years of its formation. Since the meteorite had left Mars before the serious environmental changes inimicable to life occurred, it was presumed that the life present three and one-half billion years ago had since become extinct. This scenario greatly stimulated the concept of searching for microbial fossils on Mars rather than extant life.

But Mars was not done with teasing humans. Hard on the heels of SNC ALH 84001, the analysis of another meteorite presumed from Mars, SNC EETA 79001, produced an even greater shock. This meteorite not only confirmed the biological organic evidence of the earlier one, but was estimated to be less than 600,000 years old! In terms of the planetary history of Mars, this was modern times, well after the drastic environmental changes on Mars which had been widely advertised as proof that the Viking LR could not have detected life. This news led the chemical theorists to conclude that Mars may have life today but, if so, it must be hidden in very rare "oases" deep beneath the surface of the planet in pools of liquid water heated by volcanic activity and kept from subliming into vapor by the overlying strata.

No one has offered to explain how the meteor that impacted Mars to launch EETA 79001 on its long journey to Earth found a precious oasis! Nor have the oasis proponents addressed a recent study which finds that material ejected from a planet by meteoric impact comes from near the surface not from the depths proposed for such oases.... And the leaders of the fossil search press on with no (announced) thought that might link the recent findings to the LR results! Darwin must be spinning in his grave at the thought that the modern descendants of his discipline believe that life existing on Mars during its recent era would not have survived to the present. However, they do not deny that early Earth life was faced with a much more desperate situation. When oxygen first appeared, produced by photosynthetic organisms, this gas was intensely toxic to all other life forms. However, life managed to convert that adversity to advantage.

9. Expansion of the Life Envelope

While the direct investigation of life on Mars has been at a standstill for the twenty years since *Viking*, knowledge about life on Earth has increased to an astonishing degree. Microorganisms, large anchored tubular worms, and fish, all previously unknown, have been found living in deep ocean trenches in lightless waters at temperatures of several hundred degrees Celsius and under pressures of thousands of pounds per square inch. Microorganisms frozen deep below the surface for millions of years have been resuscitated instantly when brought to the surface. Microorganisms have been found growing in cooling waters irradiated by nuclear reactors. A new kind of microorganism, capable of living on rock and water, has been found in deep sunless pools. Indeed, the "thin film of life" has been expanded to become a three-dimensional continuum. Obeying Darwin's principle of evolution, life on Earth has occupied virtually every environmental niche, many with extreme conditions exceeding those on the "hostile" Red Planet. Why should we expect life on Mars to have done less?

10. Panspermia

Perhaps the most important accomplishment of the analysis of the two life-bearing meteorites is their proof of the theory of Panspermia advanced by Svante Arrhenius in the nineteenth century. This Nobel Laureate in chemistry envisioned that life traveled through space, inoculating planet after planet. Whether the two SNC meteorites come from Mars, or not, matters little in this case. What matters most is that they do bear evidence of biology from someplace other than Earth! This finding defeats the ultimate argument of those opposing acceptance of the LR data—that the origin of life is such a complex process, still not nearly understood, that to suppose it happened on Mars is the most far-fetched explanation of the LR data possible.

"Ockham's razor," the fourteenth-century philosopher's admonition to seek no further than the simplest explanation—long cited by the pro-chemist group against the LR's having detected life—now cuts the other way! The meteorites show that we no longer have to assume that life on Mars arose there! Life, at least microorganisms, can ride the Cosmos. The way to preserve microorganisms indefinitely (no time limit is yet known, but it exceeds millions of years) is to freeze them. In the laboratory we freeze and dry them. They are readily resuscitated when placed back into an environment favorable to them. Space travel provides the best freeze-dry process available! So, microorganisms, once formed somewhere, can hitch rides for millions of years! Of course, while freeing up the LR data, this new information merely pushes back the problem of how life began somewhere.

II. L'Envoi!

Between 1976 and 1986, Pat Straat and I contended in published papers and oral statements that a biological interpretation of the LR results was possible. By 1986, our studies and our review of work done by others led to the statement that "more probably than not, the LR experiment on Mars discovered life." Much new information has been gleaned since then, on Mars and Earth, which requires a new assessment.

Each of the reasons supporting a non-biological interpretation of the LR Mars data has now been shown deficient. The demonstrated success of the LR in detecting microorganisms during its extensive test program with its record of no false positives can no longer be denied. New evidence, together with the review of the old, leaves the biological interpretation standing alone. The scientific process forces me to my new conclusion: the Viking LR experiment detected living microorganisms in the soil of Mars.

The conclusion that the Viking LR results and all available relevant evidence point to the existence of microorganisms in the soil of Mars raises the question of what type of microorganisms they might be. Several possibilities are evident. The LR data, the Viking images of greenish patches on the rocks, the Viking imaging system analysis of terrestrial lichen, and the known hardiness of lichen (these "pioneers of vegetation" are the first organisms to appear on newly formed bare rock, such as when Surtsey rose from the sea and cooled) make lichen a good candidate. Species on Earth are reported to survive on water obtained in vapor form, to endure Mars-like cold, and to grow on, even inside, rocks. The discovery and analysis of the Martian meteorites, if confirmed, would make it extremely likely that Earth and Mars have exchanged material frequently. As stated, space conditions are very good for the preservation of any microorganisms inside the ejecta. Since lichen are present within rocks on Earth, they, but not only they, make a good candidate for interplanetary travel. If present on Mars, lichen may be widely distributed over the planet's surface and might have been in the LR sample. The two symbiotic components of lichen are algae and fungi. They might also be widely distributed as individual species, as might a great variety of other species. It seems unlikely to me that any microbial forms would be confined to discrete "oases." Just as on Earth, life on Mars probably adapted to all of the environmental niches. As pointed out earlier, those niches on Mars are much less severe than on Earth. Even if that unlikely scenario of discrete oases were true, those oases might still have supplied living organisms to the LR. Organisms from the oases would have been extruded to the surface repetitively over time, perhaps by frost-heaving or by volcanic eruption, and would have become lyophilized (freeze-dried) by the climate. Thus preserved indefinitely, they would have been blown by the wind and eventually distributed over the surface of the planet. Such dormant organisms might have instantly begun to metabolize when given the LR's favorable environment and food. I think this hypothesis possible, but less likely than the hypothesis that life has adapted to all Martian ecological niches. In any event, the existence of life on Mars would make it likely that the LR soil would have contained a viable sample.

I believe confirmation of my new conclusion will come with additional life-detection missions to Mars, unfortunately not currently within NASA's plans. However, the possibility does exist that the refined cameras on *Pathfinder*, scheduled to land on Mars on July 4, 1997, may surprise us with images we readily recognize as colonies of lichen or other microorganisms. I anxiously await more Patch Rocks.

Religion, Philosophy, Society, and Science

Science's first steps, taken in mankind's fledgling society, were largely controlled by religion. Centuries of effort by truly persistent scientists, some at the cost of their lives, have still not completely freed science from the bonds of religion. However, as science grows in scope, magnitude, and importance to our everyday lives, new shackles have been forming around it, forged by politics and government. Over the past half-century, science has been transformed from a discipline engaged in by lone, gifted investigators to a major enterprise requiring elaborate facilities, equipment, and teams of researchers. Big Science requires big budgets. Funding for basic science, which promises no immediate financial return, is largely supplied by government or large philanthropic agencies. They rely upon peer review to select projects for funding. As happens within any large organization, leaders emerge to dominate policy. The scientists, like anyone in a supplicant situation, realize the desirability of maintaining favor to maintain funding. They are prone to direct their efforts to areas determined as priorities by the funding sources. They often tune their public pronouncements to the current policies of their supporters. To do otherwise may be to incur displeasure at the source, which, even though unintentional, could adversely affect funding. For example, once, when I was principal investigator on a government contract, the government contract officer paid me a usual visit to discuss progress. After some talk, he asked me to change the direction of my research to follow one of his ideas. I did not think it worth deviating from my proposed course of work and told him so. He then began to insist. I then said that I appreciated his input, but that I thought it best to stick to the plan that his agency had approved. Then I made the mistake of adding, "After all, I am the principal investigator and your agency is paying me to use my best judgment in pursuing this project." When annual funding time came around, I was told that my project would not be continued. I never was funded by that agency again.

I think that once NASA announced that the LR had produced "no evidence" for life, the scientists in the agency, outside scientists supported by the agency, and those looking to the agency for future funding took their cue. Consciously or even without overt intent, they coalesced behind the official opinion—even to the extent of using that patently inappropriate phrase "no evidence" to describe the LR's findings. The unlikely alternative is to believe that a large number of scientists do not know the definition of the word "evidence"!

In a December 12, 1996, CNN Headline News story on NASA's new "Origins" theme following a NASA-sponsored meeting between scientists and theologians, Vice President Gore and NASA Administrator Goldin were shown in a discussion. The Vice President and the Administrator were speculating on the impact that would be felt if life were discovered on Mars. Administrator Goldin commented that "... many Americans ... believe in God. Different manifestation. Taxpayer dollars are involved...." It thus seems evident that religious considerations influence NASA programs.

What role religion, philosophy, and sociological implications play in the acceptance of extraterrestrial life is difficult to assess, but I believe they do play a significant role. All three of these paradigms for seeking enlightenment would be seriously affected by the discovery of life beyond the Earth, which will have to be assimilated into the culture. We may face considerable resistance before that assimilation is accomplished. Apparently, the NASA Administrator believes the same.

The Next Steps

Before the analysis of the two Martian meteorites, NASA had laid out a ten-year plan for the continued exploration of Mars. None of these missions, some ten spacecraft in all, launched in pairs at two-year intervals, is scheduled to contain a life-detection experiment. Apparently, NASA still believes that *Viking* proved that there is no life on Mars, or else it doesn't want to discover life on Mars—yet. Despite President Clinton's stating that the number-one priority for NASA is to determine whether or not there is life on Mars, a statement echoed by the NASA Administrator, no change in plans has been announced.

Amidst all this renewal of the issue of life on Mars, NASA has announced that it will seek the "earliest possible mission, perhaps as early as the year 2001," to return a sample of Mars soil to Earth for detailed study. I believe this would subject our planet and its life forms to an undue hazard. A more cautious progression might be to send robotic missions to Mars in 1998 to settle the life issue and determine something about any life found. Then any samples for detailed study by human scientists should be returned, not to Earth, but to a laboratory established for the purpose either on the Moon or on the Space Laboratory. In that way the Earth would be protected until we were certain that there was no hazard in returning a sample to Earth.

As early as 1975, Biospherics completed a detailed report under NASA contract entitled "Technology for Return of Planetary Samples." A major problem in protecting the scientific integrity of the sample was pointed out but has yet to be addressed by NASA. Some means of protecting the sample against changes during the long return trip must be developed. A complex environmental chamber will have to be developed to maintain Martian atmospheric gas pressure and composition, Martian ambient temperature cycling, Martian diurnal lighting, water content in each of the phases occurring on Mars, pH, redox, and other still-undetermined parameters necessary to keep the sample pristine for its examination. AND, to preserve the sample to be able to examine it for living organisms, a whole life-support system must be provided and operated throughout the trip. Otherwise, any microorganisms present would likely use up some vital resource and be seriously impaired or DOA, thereby confusing the whole life issue. The report strongly recommends initial biohazard assessment and the development of suitable control technology. As stated above, the report urges that the sample not be returned directly to Earth, but to an off-Earth laboratory to determine any health or environmental threat.

The famous image returned to Earth from the Apollo Mission shows us how tiny and frail our planet is. We should protect it.

Future Experiments

A general rule for scientific investigations is that when a new technique begins to get answers, that technique is expanded to further the line of inquiry. The next experiments into the question of life on Mars, therefore, should begin with the LR technology. It readily lends itself to expansion. It not only can confirm the *Viking* results, but also can learn more about the life detected. This can readily be done by, first, separating its offerings of the left-handed and right-handed molecules, then varying the numbers and kinds of compounds to learn more about the metabolism of the life forms found. Environmental conditions can be controlled. Thus, the responses at different temperatures, relative humidities, amounts of water, atmospheric gases, and the like can be determined. An Automated Microbial Metabolism Laboratory (AMML), which Biospherics designed, built, and tested for NASA in the seventies, uses the extreme sensitivity of the radioisotope technique to look at the involvement of life-essential elements other than carbon, such as phosphorus, sulfur, and hydrogen, in the metabolism of microorganisms. We also proposed that, upon successful response to the LR probe, a subsequent mission test for the presence of adenosine triphosphate (ATP) in the cellular material detected. A robotic instrument to do this was built. Since ATP is the universal compound through which terrestrial life obtains its energy, this test would constitute a good comparison of Mars life to Earth life. New technologies, such as polymerase chain reaction (PCR) and nucleic acid mapping, will permit any traces of nucleic acids to be amplified to the point where they can be mapped for comparative studies.

In order to meet the deserved urgency and priority placed upon the search for life on Mars by President Clinton and Administrator Goldin, I suggest the following approaches for new missions:

1. The TEGA experiment in *Surveyor '98* be modified as I proposed to include life-detection capability.

2. The Hubble Telescope should be used to survey the surface of Mars for evidence of seasonal geographic changes in patterns and coloration. These should be correlated with surface temperature and atmospheric moisture content to investigate and, if found, elucidate the reported wave of darkening.

3. A new LR experiment and instrument should be sent to Mars on the earliest possible mission. This should include the suggested changes described in this chapter plus new ones that a study of this opportunity will undoubtedly develop.

4. The ATP and AMML tests should be incorporated into an updated automated instrument.
5. New techniques such as PCR, nucleic acid analysis, and high-resolution imaging should be developed, instrumented, and flown.
6. Follow-on experiments to determine the nature of any life found, its variety, and its environmental limits should be designed and flown.
7. Return sample missions should deposit samples on the Moon or aboard the Space Laboratory for detailed examinations. Before such samples are sent to Earth, complete assurance of their safety should be established.

Postscript

Many of the references supporting statements I have made in this chapter are listed in the Reference section of this book. I have not cluttered the chapter with individual citations. However, a formal scientific paper supporting my new conclusion about life on Mars is under preparation. I have been invited to present the paper at a symposium on "Instrumentation, Methods and Missions for the Investigation of Extraterrestrial Microorganisms" at the Annual Meeting of the International Society for Optical Engineering, scheduled for San Diego for late July to early August, 1997. I will update my story to try to convince my fellow scientists of the new conclusion I have reached. I hope I can, but if not, I will continue to take consolation in knowing that scientific progress is not a democratic process, and I will keep trying to turn the tide. Like the Ancient Mariner, I cannot resist telling the fascinating story about life on Mars.

Acknowledgments

I want to acknowledge and express my deep thanks to all the scientists, engineers, and technicians, too numerous to mention here, who have worked with me at Resources Research Incorporated, Hazleton Laboratories, Incorporated, and at Biospherics Incorporated over the nearly thirty years of development and analysis of this project. Many have been co-authors in our numerous scientific publications. All others have been acknowledged in those publications. Foremost is Dr. Patricia Ann Straat who, for ten years, was my right arm at Biospherics before, during, and after the Viking Mission. For a while, she literally lived with the project, in California, to assist in its final development, fabrication, and testing at TRW, Inc., the manufacturer of the LR instrument, and at JPL, the NASA center at which Viking was based. Prior to Dr. Straat's joining our company, Mary-Frances Thompson did a splendid job as my chief assistant and oversaw our laboratory efforts. Nor can I fail to mention the ingenuity of engineer George Perez, who reduced my concepts to hardware in the early years.

I am especially pleased to acknowledge the professional contributions of my son Ron, who accompanied me at JPL during the Viking Mission, having freshly graduated from high school. Now a Ph.D. physicist at MIT's Lincoln Laboratory, it was he who supported my contention that the Viking data demonstrated the presence of liquid water on the surface of Mars. He also was the first to cite Rayleigh scattering (commonly taught in college physics courses to explain why the sky is blue) to refute the red sky arbitrarily assigned to Mars by NASA, and he applied his capability with computers to help me in studying the colored patches on the rocks.

For his excellent help in editing this chapter, I want to thank Ryan Bliss of Biospherics.

I also want to thank the Biospherics Board of Directors for its patience in allowing me to participate, long after funding had ceased, in my "hobby" at the expense of more conventional and profitable things expected of a company president.

It is always fitting and proper to acknowledge family support. In this case, it's more than a perfunctory obligation. My wife, Karen (who also helped in editing this chapter), my sons, Ron and Henry, and my daughter, Carol (who also contributed to this editing), have lived with my daily struggles with Mars for more than two decades!

My ultimate thanks, however, despite the complaints I voice in this chapter and elsewhere, go to NASA, which provided me with the most exciting scientific adventure I could ever imagine, one that I believe is not over yet! Among the NASA officials I feel mostly indebted to are Drs. Freeman Quimby, Orr Reynolds (deceased), and Richard Young (recently deceased), who, as sequential early directors of the Exobiology program, enthusiastically supported my efforts.

VLR Re-Visited

Patricia Ann Straat, Ph.D.

I n the ten years that I worked on the Viking Labeled Release (VLR) experiment with Gil Levin, I never saw a response quite like the one obtained from the Mars sample. For years in advance of the actual flight experiment, hundreds of terrestrial soils were examined under Labeled Release (LR)-type conditions in thousands of tests, building a library of responses with which to compare the Mars result. Soils tested included those collected from a wide variety of terrains, such as Death Valley, all of which contained abundant microbial life and provided vigorous LR responses. In addition to soils teeming with microbial life, samples were tested that might reasonably be expected to be naturally sterile, such as those collected from the dry valleys of Antarctica. Some of these samples contained extremely low microbial populations, and accordingly, gave low LR responses, even lower than that obtained on Mars. Two of these samples were naturally sterile and gave negative LR results. Surtsey and Moon samples were also shown to be sterile and similarly gave negative LR results. (Surtsey is an island off the coast of Greenland formed in 1969 by eruption of a volcano along the mid-Atlantic ridge; samples were collected under sterile conditions immediately upon cooling of the newly formed island.)

The VLR flight data showed a definite and immediate positive response following the addition of nutrient to the Mars sample. Some features of the active cycle response were unique, particularly the gas absorption following a second nutrient injection seven days after the first injection, and it appeared that the active agent was no longer present at that time. The magnitude of the initial positive response was only about 10 percent that of an active terrestrial sample, although later laboratory experiments using endolithic microorganisms contained in an Antarctic rock provided LR data of similar kinetics and magnitude. The VLR flight data met criteria for existence of microbial

314

life, namely, that the positive response obtained during the active cycle was destroyed by 160 degrees Celsius heat sterilization, and the flight sterile control data essentially matched control data in our response library. Whatever the active agent, it was present at both landing sites some 4000 miles apart. One of the VLR samples tested was obtained from under a Martian rock; it gave a response similar in magnitude to that obtained from a sample on the exposed Martian surface, showing that the active agent was neither destroyed by UV light nor dependent on it.

Most provocative of all was the so-called cold sterilization data, the pre-heating of the Martian sample to approximately 50 degrees Celsius prior to adding nutrient. This mild treatment resulted in approximately ⅔ reduction in the magnitude of the positive response. Because few chemicals are known that exhibit such heat sensitivity, and because microbial life on this cold planet might be expected to show intolerance to such "elevated" temperatures, these data provided the strongest support for the life hypothesis on Mars.

Yet many scientists rejected the life interpretation for the VLR data. First, the kinetics and magnitude of the positive response could be accounted for by a simple chemical reaction involving only one of the VLR nutrient constituents. Second, the intense ultraviolet light that focused on the Martian surface suggested that highly reactive chemicals, possibly peroxides, might be present that could react with the VLR nutrient. Further, no other Viking experiments provided evidence for Martian life. The negative results of the organic analysis GCMS experiment were particularly damaging, because most scenarios for the life hypothesis include the presence of organic compounds in soil. And, finally, some experiments later conducted by others using flight-like hardware suggested that the VLR results could be attributed to outgassing of the VLR nutrient.

On the other hand, while the life hypothesis cannot be conclusively proven, neither can it be eliminated. Arguments can be, and have been, readily presented to counter each of the points raised above. Nothing precludes a life form utilizing only one of the VLR substrates, nor is it established that only one substrate, as opposed to small amounts of several of the substrates, was utilized. The

UV/peroxide hypothesis is plausible, but an extensive three-year effort to non-biologically replicate the active portion of the flight data using irradiation of Mars analogue soils or peroxides was largely unsuccessful. In this regard, it is extremely unfortunate that Gil Levin's experiment on board the Russian *MARS-96* was lost when the spacecraft failed. This experiment had been designed to detect chemical oxidants in the Mars surface material. As for the failure to detect organics in the Mars sample, this may reflect a lack of sensitivity in the GCMS experiment. And, finally, in my extensive experience with the LR nutrient, I simply do not agree that the flight data can be attributed to nutrient outgassing. The pitfalls in running flight simulation experiments are numerous, and the data supporting the outgassing theory appear fraught with technical problems.

Could life exist on Mars? The present Mars environment, although extremely harsh by terrestrial standards, does not necessarily preclude the existence of life. Terrestrial microbial life forms have adapted to a wide range of environmental conditions, including extremes of temperature, atmospheric and osmotic pressure, pH and atmospheric gases. Many terrestrial obligate anaerobes have an absolute requirement for a carbon dioxide atmosphere. Antarctic microorganisms grow at low temperatures, and certain terrestrial bacteria and invertebrates are known to survive extreme desiccation for years in a state of "cryptobiosis." These organisms can be revived upon addition of water. Of extreme interest is the report that a Streptococcus microorganism was recovered from the *Surveyor III* spacecraft by the *Apollo 12* crew. Clearly this microorganism survived the extremely harsh conditions on the surface of the Moon for several years. Survival data of this type are numerous, and suggest that certain terrestrial microorganisms could survive under Martian conditions. Hopefully, the planet has not already been contaminated by previous US and USSR spacecraft that have landed on Mars. But, surely, if survival is possible for terrestrial microorganisms under Mars or Mars-like conditions, then indigenous Martian life, which should be well-adapted to its environment, is not precluded by the harshness of the environment.

Perhaps a more fundamental question is whether life could have originated on Mars. Numerous photographs from both *Mariner 9* and the Viking Mission have provided ample evidence for liquid water in the Martian past. This in turn indicates significantly different temperature and pressure conditions in the past that may have been conducive to the formation of life. It is hypothesized that life, at least microbial life, formed early in Earth's history, perhaps as early as the formation of the first sedimentary rocks several billion years ago. The early Martian environment may not have been significantly different from that of primordial Earth, in which life presumably did originate.

The recent discovery of the Mars meteorites and possible ancient life forms contained therein, if true, is perhaps the most astounding scientific discovery of the century. If life did indeed originate on the Red Planet, evidence such as that briefly discussed above suggests that it is not totally improbable that it still exists today, at least in microbial form. If it is true that microbial life has evolved and continued to exist over millions or billions of years, then it is also not impossible that the VLR experiment actually detected it. The experiment is highly sensitive and capable of detecting as few as 10 microbes in a terrestrial sample. These are a lot of "ifs," but the data are sufficiently provocative that the possibility of having already detected life on Mars must be seriously considered.

The reluctance to accept the VLR data as possible evidence for Martian microbial life has perhaps held up the search for twenty years. Had the popular interpretation of the provocative VLR data left open the possibility of life on Mars, the search may have continued. Follow-on Viking missions had been in the planning stages even in the early seventies. Yet in the recent sudden surge of interest generated by the Martian meteorites, we must be extremely careful in our approach. We must not contaminate the planet, and we must not assume that Mars is sterile or that any indigenous life is harmless to terrestrial life forms. I cannot over-emphasize the need for caution in considering a Return Mars Sample mission. While sophisticated instruments here on Earth would greatly facilitate the study of a Martian sample, the probability of life on Mars is sufficiently high to warrant caution and not bring unknown creatures back to Earth until it is well established that it is safe to do so. In the interim, studies could be conducted remotely, if not on Mars, then on the Moon, or on a space station.

317

But proceed we must! This is perhaps the most exciting venture of our times, and it is within our capabilities to explore this question. If life on Mars is actually proven not only to exist, but to have arisen independently, this would be an extraordinary discovery with major implications. If it ever evolved, it seems likely that it is still there, well adapted to its environment. Did the VLR experiment discover life on Mars? I only hope the truth is known in my lifetime!

References

Introduction

Billingham, J., ed. 1981. Life in the Universe. NASA CP-2156, Washington, D.C.

Cameron, A. G. W. 1962. The Formation of the Sun and Planets. *Icarus* 1: 13.

Cameron, A. G. W. 1975. The Origin and Evolution of the Solar System. *Scientific American*, September, p. 32.

Donn, B. 1978. Condensation Processes and the Formation of Cosmic Grains. In *Protostars and Planets*, ed. T. Gehrels and M. Matthews. Tucson: University of Arizona Press.

Haldane, J. 1929. *The Origin of Life*. Also:1932. The Inequality of Man. London: Chatto and Windys.

Hartman, H.; Lawless, J. G.; and Morrison, P., eds. 1985. Search for the Universal Ancestors. NASA SP-477, Washington, D.C.

Levin, G. V., and P. A. Straat. 1981. Antarctic Soil No. 726 and Implications for the Viking Labeled Release Experiment. *J. Theo. Biol.*, v. 91, p. 41.

Lewis, J. S., and R. G. Prinn. 1984. *Planets and Their Atmospheres*. New York: Academic Press.

Miller, S. 1953. *Science*, v. 117, p. 528.

Miller, S., and Harold C. Urey. 1959. *Science*, v. 130, p. 245.

Oparin, A., 1924. *Proiskhozhdeniye zhizni (The Origin of Life)*. Moscow: Izdatel'stvo "Moskovsky Rabochiy."

The Scientific Results of the Viking Project. 1977. Washington, D.C: American Geophysical Union. [Author's note: This collection of reprints from the *Journal of Geophysical Research* should be in the library of every serious student of Mars. Copies are still available from the AGU.]

Chapter 1

Brock, T. D. 1961. *Milestones in Microbiology*. New Jersey: Prentice-Hall.

Brugsch. 1891. *Religion and Mythology of the Ancient Egyptians*. Germany: Leipzig.

Dobell, C. 1932. *Antony van Leeuwenhoek and His Little Animals*. London: Russell and Russell.

Hoyt, W. G. 1976. *Lowell and Mars*. Arizona: University of Arizona Press.

King, H. C. 1979. *The History of The Telescope*. New York: Dover.

Ley, W. 1963. *Watchers of the Skies*. New York: Viking Press.

Lowell, P. 1906. *Mars and Its Canals*. New York: Macmillan.

Oparin, A. I. 1962. *Life: Its Nature, Origin and Development*. New York: Academic Press.

Richardson, R. S., and C. Bonestell. 1964. *MARS*. New York: Harcourt, Brace & World, Inc.

Sheehan, W. 1988. *Planets and Perception*. Arizona: University of Arizona Press.

Shklovskii, I. S., and C. Sagan. 1966. *Intelligent Life in the Universe*. New York: Holden-Day.

Thomson, W. (Lord Kelvin, W.). 1871. *Brit. Assoc. Reports*, v. 41, p. 84.

Chapter 2

Allamandola, L. J., and S. A. Sanford. 1988. *Dust in the Universe*. New York: Cambridge University Press.

Blanchard, D. C., and Woodcock, A. H. 1957. Bubble formation and modification in the sea and its meteorological significance. *Tellus*, v. 9, no. 2, pp. 145–158.

Clark, B. C. 1979. Is the Martian Lithosphere Sulfur Rich? *J. Geophys. Res.*, v. 84, p. 8395.

Focas, J. H. 1962. Seasonal Evolution of the Fine Structure of the Dark Areas of Mars. *Planetary and Space Science*, v. 9, p. 371.

Glasstone, S. 1968. *The Book of Mars.* NASA SP-179, Washington, D.C.

Inhabited Space, 1975. NASA report TT-F-819.

Krumbein, W. E. 1968. *Recent Developments in Carbonate Sedimentology in Central Europe.* New York: Springer-Verlag.

Latarjet, R. 1952. C.R. *Acad. Sci.*, p. 235.

Lechevalier, H. A., and M. Solotorovsky. 1965. *Three Centuries of Microbiology.* New York: McGraw-Hill.

MacIntyre, Ferren. 1974. Chemical fractionation and sea-surface micro-layer processes, Chap. 8 *of* Goldberg, E. D., ed., *The Sea*, v. 5, *Marine Chemistry*, New York: Wiley.

Miyake, Yasuo, and Tsunogai, Shizuo. 1963. Evaporation of iodine from the ocean. *J. Geophys. Res.*, v. 68, no. 13, pp. 3989–3993.

Moore, H. J. 1977. *The Scientific Results of The Viking Project.* Washington, D.C., American Geophysical Union.

Opik, E. J. 1966. The Martian Surface. *Science*, v. 153, p. 255.

Pavlovskaya, T. E., and T. A. Telegina. 1979. Photochemical conversions of lower aldehydes in aqueous solutions and in fog. *Origins of Life*, p. 303.

Pflug, H. D. 1986. Morphological and chemical record of organic particles in precambrian sediments. *Syst. Appl. Microbiol.*, v. 7, p. 184.

Rea, D. G., and B. T. O'Leary. 1965. Visible Polarization data of Mars, *Nature*, v. 206, p. 1138.

Sakugawa, H., and I. Kaplan. 1990. Atmospheric Hydrogen Peroxide. *Environ. Sci. technol.*, v. 24, no. 10.

Salisbury, F. B. 1962. Martian Biology. *Science*, v. 136, p. 17.

Sanford, S. A., and Y. Pendleton. 1991. *Astrophs. J.*, v. 371, p. 607.

Schidlowski, M. 1983. Biology mediated isotope fractionations: biochemistry, geochemical significance and preservation in earth's oldest sediments. Dordrecht (the Netherlands).

Schonbein, C. F. 1855. Verhandl. naturforsch. Ges. Basel, I, 339.

Schwabe, A. 1960a. Blaualgen aus ariden Boden, Forsch. *u Fortschr.*, v. 34, pp. 194–197.

Sigg, A., and A. Neftel. 1991. Evidence for a 50% increase in hydrogen peroxide over the past 200 years from a Greenland ice core. *Nature*, v. 351, pp. 557.

Sinton, W. M. 1957. Spectroscopic Evidence of Vegetation on Mars. *Astrophs. J.*, v. 126, p. 231.

Sinton, W. M. 1959. Further Evidence of Vegetation on Mars. *Science*, v. 130, p. 1234.

Sinton, W. M. 1961. Identification of Aldehydes in Vegetation Regions of Mars. *Science*, v. 134, p. 529.

Slipher, E. C. 1962. *The Photographic Story of Mars.* New Jersey: Sky Publishing Corp.

Sturges, W. T. 1992. Surface ozone depletion in Arctic spring sustained by bromine reactions on aerosols. *Nature*, v. 358, p. 552.

Sturges, W. T., 1992. Bromoform emission from Arctic ice algae. *Nature*, v. 358, p. 660.

The Martian Landscape. 1978. NASA SP-425. Washington, D.C.

Thorpe, T. E. 1979. A History of Mars Atmospheric Opacity in the Southern Hemisphere During the Viking Extended Mission. *J. Geophys. Res.*, v. 84, p. 6663.

Tikhov, G. A. 1972. *Naselennyy Kosmos.* Moscow: Nauka Press.

Wickramasinghe, D. T., and D. A. Allan. 1980. The 3.4 micron interstellar absorption feature. *Nature*, v. 287, p. 518.

Winogradsky, S. N. 1887. Concerning sulfur bacteria. *Botanische Zeitung*, v. 45, p. 489 (in German).

Zika, R. E., and W. L. Saltzman. 1982. Hydrogen peroxide levels in rainwater collected in South Florida and the Bahama Islands. *J. Geophys. Res.*, v. 84, p. 5015.

Chapter 3

Alexander, M. 1969. *Nature*, v. 222, p. 432.

Benoit, R. E., and C. E. Hall. 1970. *Antarctic Ecology.* (M.W. Holgate, ed.) New York: Academic Press.

Cameron, R. E. 1971. *Research in the Antarctic.* (L. O. Quam and H. D. Porter, eds.), Publ. No. 93. Washington D.C. American Association for the Advancement of Science, Washington, D.C.

Chikahiro, M., Michihata, F., and K. Asada, 1991. Scavenging of Hydrogen Peroxide in Prokaryotic and Eukaryotic Algae. *Plant Cell Physiol.*, v. 32, p. 33.

Davies, R. W., and M. G. Comuntzis. 1959. The Sterilization of Space Vehicles to Prevent Extraterrestrial Biological Contamination. London: Proceedings, Tenth International Astronautical Congress.

Davies, R., and A. J. Sinskey. 1973. *J. Bacteriol.*, v. 113, p. 133.

Glasstone, S. 1968. *The Book of Mars.* NASA SP-179, Washington, D.C.

Horowitz, N. H., and others, 1969. Sterile Soil from Antarctica: organic analysis. *Science*, v. 164, p. 1054.

Horowitz, N. H., Cameron, R. E., and J. S. Hubbard. 1972. *Science*, v. 176, p. 242.

Hubbard, J. S., Hobby, G. L., Horowitz, N. H., and others. 1970. Measurement of Carbon Dioxide Assimilation in Soils: an Experiment for the Biological Exploration of Mars. *Appl. Microbiol.*, v. 19, p. 32.

Levin, G. V., Heim, A. H., Clendenning, J. R., and M. F. Thompson. 1962. Gulliver—A Quest for Life on Mars. *Science*, v. 138, p. 114.

Levin, G. V. 1965. Significance and Status of Exobiology. *Bioscience*, v. 15, p. 17.

Levin, G. L., and A. H. Heim. 1965. Gulliver and Diogenes—Exobiological antitheses. (COSPAR), Life Sciences and Space Research III, p. 105, (Amsterdam).

The Scientific Results of the Viking Project. 1977. Washington, D.C.: American Geophysical Union.

Vishniac, W. V., 1960. Extraterrestrial Microbiology. *Aerospace Medicine*, v. 31, p. 678.

Vishniac, W. V., and S. E. Mainzer. 1972. Soil Microbiology Studied in situ in the Dry Valleys of Antarctica. *Antarctic Journal of the United States*, v. 7, p. 88.

Chapter 4

Biemann, K., and others. 1977. The Search for Organic Substances and Inorganic Volatile Compounds in the Surface of Mars. *J. Geophys. Res.*, v. 82, p. 4641.

Cooper, H.S.F., Jr. 1976. *The Search For Life On Mars.* New York: Holt, Rinehart and Winston.

Davis, I., and J. D. Fulton. 1960. The Reactions of Terrestrial Microorganisms to Simulated Martian Conditions. Proceedings, Tenth International Astronautical Congress, p. 778 (London).

Edwards, J. O. 1962. *Peroxide Reaction Mechanisms.* New York: Wiley.

Ezell, E. C., and L. N. Ezell. 1984. *ON MARS.* NASA SP-4212, Washington, D.C.

Horowitz, N. H., Hobby, G. L., and J. S. Hubbard. 1977. Viking On Mars: The Carbon Assimilation Experiments. *J. Geophys. Res.*, v. 82, p. 4659.

Hunten, D. M. 1979. Possible oxidant sources in the atmosphere of Mars. *J. Mol. Evol.*, v. 14, p. 71.

Lavoie, J. M., Jr. 1979. Support Experiments to the Pyrolysis/Gas Chromatographic/Mass Spectrometer Analysis of the Surface of Mars. Ph.D. Dissertation. Massachusetts Institute of Technology, Cambridge, Mass.

Levin, G. V., and P. A. Straat. 1979. Completion of the Viking Labeled Release Experiment on Mars. *J. Mol. Evol.*, v. 14, p. 167.

Levin, G. V., and P. A. Straat. 1980. Development of Biological and Nonbiological Explanations for the Viking Labeled Release Data. Final Report to NASA, Contract NASW-3249, Washington, D.C.

Mehler, A. H. 1951. Studies on reactions of illuminated chloroplasts 1. *Arch. Biochem. Biophys.*, v. 3, pp. 65–77.

Moore, H. J., and others. 1978. Rock Pushing and Sampling Under Rocks on Mars. U.S. Geological Survey Professional Paper 1081. Washington, D.C.

Moore, H. J., and others. 1987. *Physical Properties of the Surface Materials at the Viking Landing Sites on Mars.* U.S. Geological Survey Professional Paper 1389. Washington, D.C.

Oyama, V. I., and B. I. Berdahl. 1977. The Viking Gas Exchange Experiment Results from Chryse and Utopia Surface Samples. *J. Geophys. Res.*, v. 82, p. 4669.

The Scientific Results of the Viking Project. 1977. Washington, D.C.: American Geophysical Union.

Van Baalen, C. 1965. Quantitative surface plating of coccoid blue-green algae. *J. Phycol.*, v. 1, pp. 19–22.

Van Baalen, C., and J. E. Marler. 1966. Occurrence of hydrogen per-oxide in seawater. *Nature*, v. 211, p. 951.

Chapter 5

Arvidson to DiGregorio, personal communication, 1995.

Averner, M. M., and R. D. MacElroy, eds. 1976. *On the Habitability of Mars.* NASA SP-414, Washington, D.C.

Banin, A., and J. Rishpon. 1979. Smectite clays in Mars soils: Evidence for their presence and role in Viking Biology results. *J. Mol. Evol.*, v. 14, p. 133.

Beimann to DiGregorio, personal communication, 1996.

Guinness, E., Arvidson, R., and others. 1979. Reports of the Planetary Geology Program. NASA Technical Memorandum—80339. Washington, D.C.

Hubble Space Science Institute. 1991. Hubble Space Telescope to Monitor Changes on Mars. News release No. 91-05. Maryland.

Hubble Space Science Institute. 1994. Three Color Composite/Wide Field Planetary Camera. Paper STScI-91-11. Maryland.

Hubble Space Telescope News, March 21, 1995, Photo Release No. STScI-PRC95-17B. Maryland.

James, P. B., Clancy, R. T., Lee, S. W., Martin, L. J., and others. 1994. Monitoring Mars with the Hubble Space Telescope: 1990–1991 Observations. *Icarus*, v. 109, p. 79.

Levin, G. V., and P. A. Straat. 1978. Color and Feature Changes at Mars Viking Lander Site. *J. Theor. Biol.*, v. 75, p. 381.

Levin, G. V., and P. A. Straat. 1981. A Search for a Nonbiological Explanation of the Viking Labeled Release Life-Detection Experiment. *Icarus*, v. 45, p. 494.

Levin, G. V., and P. A. Straat. 1981. Antarctic Soil No. 726 and Implications for the Viking Labeled Release Experiment. *J. Theor. Biol.*, v. 91, p. 41.

Strickland to DiGregorio, personal communication, 1996.

Strickland, E. L., III. 1979. Reports of the Planetary Geology Program 1978–1979. NASA Technical Memorandum—80339. Washington, D.C.

Strickland, E. L., III. 1981. Recent Weathering of Rocks at the Viking Landing Sites: Evidence from Enhanced Images and Spectral Estimate Ratios. Third International Colloquium on Mars, LPI Contribution 441, p. 253.

Strickland, E. L., III. 1981. Seasonal and Secular Changes of Martian Albedo Patterns: Analysis of Airbrushed Albedo Maps. Third International Colloquium on Mars, LPI Contribution 441, p. 256.

Strickland, E. L., III. 1986. Color/Albedo Distribution of Soil and Rock Surfaces at the Viking 1 Landing Site. *Lunar and Planetary Science XVII*, v. 2, p. 835.

The Scientific Results of the Viking Project. 1977. Washington, D.C. American Geophysical Union.

Chapter 6

DeVincenzi, D. L., H. P. Klein, and Bagby. 1991. Planetary Protection Issues and Future Mars Missions. NASA Ames Research Center, Moffett Field, California.

Friedmann, E. I. 1980. Endolithic Microbial Life in Hot and Cold Deserts. *Origins of Life*, v. 10, p. 223.

Levin, G. V. 1991. Life-Detection Experiment For MARS-94 Lander—Abstract Of Proposal. Biospherics Incorporated, Maryland.

Levin, G. V. 1992. Proposal for: Investigation of Active Agent in Mars Surface Material. Biospherics Incorporated, Maryland.

Levin, G. V. 1995. Technical Proposal: Analysis: Life or Inorganic Forms—ALIF. Biospherics Incorporated, Maryland.

Mitchell, F. J., and W. L. Ellis. 1971. Surveyor III: Bacterium isolated from lunar retrieved TV camera. Proceedings of the Second

Lunar Science Conference, v. 3, p. 2721. MIT: Massachusetts.

NASA. 1980. Standard Procedures for Microbiological Examination of Space Hardware. NHB 5340.1B. NASA, Washington, D.C.

NASA. 1990. Lessons Learned from the Viking Planetary Quarantine and Contamination Control Experience. NASA Contract Document No. NASW-4355. NASA, Washington, D.C.

Robinson, G. S. 1991. Exobiology and Planetary Protection—The Evolving Law. An unpublished paper presented to the Space Studies Board Planetary Protection Workshop, NAS Beckman Center, September 13. Washington, D.C.

Space Sciences Board. 1976. On Contamination of the Outer Planets by Earth Organisms. Report of the Ad Hoc Committee on Biological Contamination of Outer Planets and Satellites, Panel on Exobiology, March 20. National Academy of Sciences, Washington, D.C.

Weber, P., and J. M. Greenberg. 1985. Can Spores Survive in Interstellar Space? *Nature*, v. 316, p. 403.

Zvyagintsev, D. G., and others. 1985. Survival time of microorganisms in permanently frozen sedimentary rocks and buried soils. (in Russian) *Mikrobiologiya*, v. 54, p. 155.

Chapter 7

Alexander, M. 1971. *Microbial Ecology*. New York: Wiley.

Booth, N. 1996. The Secret of Mars. article for *The London Times*, 11/11/96.

Chandler, D. 1996. Seeking Life Beyond Earth: the rush to Mars is on. article for *The Boston Globe*, 11/04/96.

Goldsmith, D., and T. Owen. 1979. *The Search for Life in the Universe*. Menlo Park: Benjamin/Cummings.

Gould, S. J. 1974. The Great Dying. *Natural History*, v. 83, p. 22.

Hoyle, F., and N. C. Wickramasinghe. 1978. *Lifecloud*. New York: Harper & Row.

Hoyle, F. 1984. *The Intelligent Universe*. New York: Holt, Rinehart and Winston.

Levin, G. V. 1996. Technical Proposal: Identification of Mars Soil

Active Agent by Enhancement of *Surveyor '98* TEGA Experiment. Biospherics Incorporated, Maryland.

Levin, G. V. 1996. Re-examine Evidence of Life. Editorial in: *Space News*, October 14–20.

Underwood, J. 1975. *Biocultural Interactions and Human Variations.* Dubuque, Iowa: W.C. Brown.

Chapter 8

Alexander, M., and others. 1980. *American Type Culture Collection Methods I. Laboratory Manual on Preservation, Freezing and Freeze-Drying as Applied to Algae, Bacteria, Fungi and Protozoa.* American Type Culture Collection, Rockville, Maryland.

Booth, W. E. 1941. Algae as pioneers in plant succession and their importance in erosion control. *Ecology,* v. 22, p. 38.

Bridges, B.A. 1976. Survival of Bacteria Following Exposure to Ultraviolet and Ionizing Radiations. *Sym. Soc. Gen. Microbiol.,* v. 26, p. 183.

Clark, B. C., and D. C. van Hart. 1981. The Salts of Mars. *Icarus,* v. 45, p. 684.

Fliermans, C., and D. Balkwill. 1989. Microbial Life in Deep Terrestrial Subsurfaces. *BioScience,* v. 39, p. 370.

Friedmann, E. I. 1993. Extreme environments and exobiology. *Giornale Botanico Italiano,* v. 127, p. 369.

Friedmann, E. I., Hua, M., and R. Ocampo-Friedmann. 1993. Terraforming Mars: Dissolution of carbonate rocks by cyanobacteria. *Journal of The British Interplanetary Society,* v. 46, p. 291.

Hess, S. L., and others. 1977. Meteorological Results from the surface of Mars: Viking 1 and 2. *J. Geophys. Res.,* v. 82, p. 4559.

Hoham, R. W. 1989. Snow as a habitat for microorganisms. *Exobiology and Future Mars Missions.* NASA CP-10027, p. 32. Washington, D.C.

Kapitsa, A. P., and others. 1996. A large deep freshwater lake beneath the ice of central East Antarctica. *Nature,* v. 381, p. 684.

Krumbein, W. E. 1968. *Recent developments in carbonate sedimentology in central Europe.* New York: Springer-Verlag.

McKay, C. P., Friedmann, E. I., and others. 1992. History of Water on Mars: A Biological Perspective. *Adv. Space Res.*, v. 12, p. 231.

McKay, D. S., Gibson, E. K., Jr., Thomas-Keprta, K. L., Zare, Z. N., and others. 1996. Search for Past Life on Mars: Possible Relic Biogenic Activity in Martian Meteorite ALH 84001. *Science*, v. 273, p. 924.

Norton, C. F., and W. D. Grant. 1988. Survival of halobacteria within fluid inclusions in salt crystals. *J. Gen. Microbiol.*, v. 134, p. 1365.

Phelps, T. J., and others. 1992. Comparison of Geochemical and Biological Estimates of Subsurface Microbial Activity. *Appl. Environ. Microbiol.*, v. 58.

Rundel, P. W., and others. 1991. The Phytogeography and Ecology of the Coastal Atacama and Peruvian Deserts. *Alisio*, v. 13, p. 1.

Sprague, A. L., Hunten, D. M., and others. 1996. Martian water vapor 1988–1995. *J. Geophys. Res.*, v. 101, pp. 23, 229.

Stevens, T. O., and J. P. McKinley. 1995. Lithoautotrophic Microbial Ecosystems in Deep Basalt Aquifers. *Science*, v. 270, p. 450.

Further Reading

I would like to point out to the readers of this book who wish to find out more about the Mars life issue the following suggested reading material. While some of what is contained in these documents is technically oriented, there is enough "plain language" for the educated layman to understand the significance of each.

Mars Sample Return: Issues and Recommendations—a report published by the National Academy Press on March 6, 1997. Copies can be obtained free by writing to:

Space Studies Board
National Research Council
2101 Constitution Avenue NW
Washington, DC 20418

Since the time of the Viking Biology experiments on the surface of Mars beginning in 1976 until March 6, 1997, the NASA scientific community has maintained that the Martian surface is effectively sterilizing for all forms of terrestrial microorganisms. Their reasoning for more than twenty years has been based on the following flawed concepts addressed extensively throughout *Mars: The Living Planet* and briefly reviewed again here:

Low pressure. Because the atmospheric pressure on Mars varies from 7.4 to 10 millibars (extremely low by terrestrial standards), it would damage any organisms and affect efficient DNA repair mechanisms.

Low temperature. Any organisms on Mars would freeze and their cells would be damaged or killed.

Water. Liquid water under current Martian atmospheric pressure is unstable, and dry conditions would lead to mutations and organism death.

Radiation. Ultraviolet radiation from the Sun impacting Mars without the protection of a substantial ozone layer would obliterate any surface organisms.

Oxidants. Hypothesized substances NASA claims are produced in the atmosphere (hydrogen peroxide) with others formed at the surface by the interaction of UV radiation and the topmost layer of Martian soil would kill any life.

Carbon dioxide. Believed by some NASA scientists to cause low pH conditions that would be damaging to cellular proteins and cell metabolism.

No organic material. Because of UV radiation and the strong oxidants NASA scientists claim are in the soil, no organic material would accumulate on the surface and thus there would be no "food" for organisms on or near the surface.

With the publication of this report in 1997 by the Space Studies Board, we read what appears to be a changing perspective about life on Mars:

> each returned sample should be assumed to contain viable exogenous biological entities until proven otherwise.... Contamination of the Earth by putative Martian microorganims is unlikely to pose a significant ecological impact or other significant harmful effects ... the risk is not zero, however.

Nowhere in this volume does it mention the importance of further biological testing on the surface of Mars before bringing a sample to Earth. Why? If the risk is not zero, even if there is a one-in-a-million chance, then why bring it back if it could possibly alter our biosphere? *Mars Sample Return: Issues and Recommendations* appears to be another NASA smoke-screen justifying the bringing back of a sample from Mars. Nonetheless, it is an interesting document that clearly shows the careful sidestepping of the issue of sending other biology detection instruments to Mars, as you have read in this book.

Another important document appears in the March 25, 1997, issue of the *Journal of Geophysical Research*, volume 102, pages 6525–6534, entitled "High-resolution spectroscopy of Mars..." by V.A. Krasnopolsky, G.L. Bjoraker, M.J. Mumma, and D.E. Jennings. All these scientists are from the Laboratory for Extraterrestrial Physics at the NASA Goddard Space Flight Center. One of the authors of this extremely important paper, Dr. Michael J. Mumma, is Chief Scientist for Extraterrestrial Physics at NASA.

The actual research for this paper was conducted in 1988 using the 4-m telescope at Kitt Peak National Observatory in Arizona. Why did it take so long (1997) for the study to make it into print? One of the reasons might be because Krasnopolsky, Bjoraker, Mumma, and Jennings did the most sensitive Earth-based search ever attempted for hydrogen peroxide in the atmosphere of Mars and did not detect any hydrogen peroxide whatsoever. These scientists conclude "...the abundance of hydrogen peroxide is also an important factor in interpretations of the Viking Labeled Release life science experiment because peroxides in soil particles have been suggested as an alternative explanation for the positive response seen in that instrument...." As Dr. Levin would say, "The evidence mounts."

The entire issue of the February 25, 1997, *Journal of Geophysical Research*, volume 102, is devoted to the Mars Pathfinder Mission. I recommend this special issue to readers because it contains important information about the *Mars Pathfinder* color cameras that will be on board both the microrover and *Pathfinder Lander*. Obviously, high-resolution color imaging of the Martian surface could serve to discover conclusively whether lifeforms similar to algae, bacteria, lichens, or cyanobacteria exist at the surface of Mars or on the rocks. However, reading through this issue we find that the microrover will have three cameras: two monochrome (black-and-white) and one aft color camera.

If you recall the discussion in this book on how NASA used the *Viking Lander* camera color infrared diodes to color Mars a uniform orange-red, we (apparently) are going to be given similar views by the microrover. Instead of a camera comprised of red, green, and blue filters capable of showing "true colors," the *Pathfinder Rover* has green, red, and infrared filters, almost ensuring a "red planet."

The *Mars Pathfinder Lander* is another story. According to the information contained in the Mars Pathfinder special issue, page 4011, under the sub-heading "Color Imaging" we read that the Lander camera has a color stereo imaging capability. One "eye" of this camera will make images from red, green, and blue filters—true color imaging. However, the other part (eye) of this color stereo imaging system has only a red and blue filter in combination with an infrared component. Simply stated, if the red, green, and blue images from the one *Pathfinder Lander* camera eye—the one capable of producing true colors—are released without the infrared component of the second camera eye, then we will see the true colors of Mars. If not, be prepared to view orange-red tweaked images reminiscent of what we viewed with the Viking Landers in 1976.

The *Journal of Geophysical Research* can be found at any college or university science library, or back issues can be ordered by calling the American Geophysical Union at (202) 462-6900.

Finally, the important issue of whether the SNC Martian meteorites contain terrestrial contaminants from Antarctica instead of evidence of Martian biota has been an interesting debate. The flurry of both pro and con scientific papers has been fast and furious. For those who would like to better understand this important scientific debate via a highly readable format intended for the educated layman, I recommend Dr. Allan Treiman's excellent InterNet Site located at the Lunar and Planetary Institute of the Johnson Space Center. For those with access to a computer connected to the InterNet the number is: http://cass.jsc.nasa.gov/lpi/meteorites/marsunderscoremeteorite.html.

The name of Dr. Treiman's excellent on-going discussion is "Recent Scientific Papers on ALH 84001 Explained with Insightful and Totally Objective Commentaries." Enjoy!

—Barry E. DiGregorio

Glossary

aerobe: An organism capable of growing in the presence of oxygen.

aerobic: The utilization of oxygen during respiration.

aerosol: Atomized particles suspended in an atmosphere.

aldehyde: An organic compound that can be formed in some primitive planetary atmospheres and deposited on the surface through the process involving photodissociation (solar ultraviolet light).

alga: Photosynthetic unicellular or multicellular organisms (algae). Some of the bacterial algae are faculative aerobes and anaerobes, being able to use oxygen when it is available or carbon dioxide when it is not. The blue-green algae can live under a pure carbon dioxide atmosphere and are differentiated from higher plants by their lack of roots, stems, or leaves. Algae contain chlorophyll.

amino acid: A basic organic constituent of a protein molecule necessary for terrestrial life. Amino acids have been found in Martian meteorite EETA 79001.

anaerobe: A microorganism that does not use oxygen to obtain energy or grow. Oxygen is toxic to these organisms and most use carbon dioxide exclusively.

autotroph: An organism that manufactures organic nutrients from inorganic substances, such as rocks and minerals.

bacillus: Any rod-shaped bacterium.

bacteriochlorophyll: A photosynthetic pigment used by a variety of anaerobic photosynthetic bacteria.

bacteriophage: A group of viruses that infect, parasitize, and kill bacteria.

bacterium: A extremely small, typically unicellular organism characterized by a lack of a nucleus.

basalt: A dark-colored, fine-grained volcanic rock found on the Moon, Earth, and Mars.

biochemistry: A science dealing with the chemistry of living organisms.

bioluminescence: The chemically produced emission of light by living organisms.

biomass: The mass of living matter in a particular area.

biome: A habitat zone such as a desert or tundra.

biosphere: The zone of a planet that includes the lower atmosphere and layers of soil and water where living organisms are found.

biota: A community of organisms in a given region.

blue sky photograph: The first color *Viking Lander 1* image of Mars returned to Earth in 1976, which showed the surface of Mars with a brown and reddish soil with green spots on rocks and a blue sky color. The image was ordered to be destroyed by the NASA Administrator, Dr. James Fletcher.

calcrete: Almost any terrestrial material which has been cemented and/or replaced by dominantly calcium carbonate. In desert soils the method of calcification can be hydrologic, geological, and biochemical reactions or combinations of all three processes. Calcrete covers approximately 13 percent of the Earth's land surface where precipitation is less than 600 mm annually.

caliche: An essentially calcium carbonate deposit (up to 98 percent) remaining in or on soil following loss of water through evaporation. Biochemical actions include the chelating of calcium generated by algae. Lichen, bacteria, and fungi all play a role in the development of desert caliches. In 1976, the *Viking Landers* found much evidence for something similar to caliche in the soil of Mars.

carbon dioxide: A colorless gas readily dissolved in water. Over 90 percent of the atmospheric carbon dioxide on Earth today is produced by bacteria. Mars has a 95 percent carbon dioxide atmosphere.

carbonate rocks: Rock rich in calcium carbonate that can be formed through a process involving dissolved carbon dioxide in water and colonies of bacteria, and also various other marine organisms.

carbonic anhydrase: An enzyme that catalyzes the reversible conversion of carbonic acid to carbon dioxide gas and water.

carotenoid: A pigment sometimes produced to protect an organism from damaging ultraviolet light.

catalase: An enzyme that can be excreted by microorganisms to convert hydrogen peroxide into water and oxygen.

chemoautotroph: An organism that obtains energy by oxidizing inorganic compounds such as sulfur and iron, using carbon dioxide as its only source of carbon.

chemolithoautotroph: Anaerobic bacteria that grow on hydrogen that is produced from inorganic reactions between rock and water.

chemolithotroph: Organisms that manufacture food with the aid of energy obtained from chemicals and with inorganic raw materials contained in rocks, e.g., sulfur and iron.

chemotroph: An organism that lives on the energy of inorganic chemical reactions.

chlorophyll: A light-energy-trapping green pigment used by green plants for photosynthesis.

colony: A mass of microorganisms growing together which have cells that share a common origin.

cyanobacteria: Any large group of blue-green photosynthetic bacteria having as photopigments chlorophyll, phycocyanin, and phycoerythrin and producing oxygen as a photosynthetic waste product. They are considered to be the most ancient surviving aerobic-photosynthetic organisms on the Earth.

Deinococcus radiodurans: A highly tolerant strain of radiation-resistant bacteria sometimes found living on the interiors of nuclear reactors. They survive by constantly repairing their DNA with a specialized biologically produced enzyme. Ionizing radiations of 5,000 to 30,000 megarads are often tolerated by these microorganisms. By comparison, a lethal dose of ionizing radiation for a human is 100 rads.

desert crust: Sometimes referred to as "Nari-lime-crust," first found in Turonian limestone on the deserts of Israel. It is a hard calcareous crust containing large amounts of autotrophic and heterotrophic bacteria, fungi, actinomycetae, and green and blue-green algae. Desert crust is known to form the ground cover in areas measuring thousands of square meters in Utah and Colorado. This algal mat showing stromatolithic features such as sediment trapping and accretion might have been found on Mars by the *Viking Landers.*

desert varnish: Forms as a result of geochemical and biochemical weathering processes involving transportation and deposition of manganese or iron oxide and blue-green algae. It is believed that dew which forms in the desert after dusk provides the thin film of moisture necessary for the algae's survival. This characteristic dark but shiny coating observed on terrestrial desert rocks can also be seen on some Martian rocks in the *Viking Lander* images.

desiccate: To remove water.

dissociate: A process of breaking apart molecules through the use of ultraviolet light (photodissociation) or chemical energy. On Mars, the process of photodissociation is proposed as a possible mechanism for the breaking up by solar UV of Martian water molecules into hydrogen peroxide. However, to date no evidence of hydrogen peroxide has ever been found on Mars.

ecology: The study of the interrelationships that exist between organisms and their environments.

ecosystem: The combined systems of organisms and their environments functioning as a coherent whole.

enzyme: A biological catalyst (protein) produced by an organism.

exoenzyme: An enzyme excreted by a microorganism into its environment.

exospore: A desiccation- and heat-resistant spore shell formed externally by a microorganism.

exotoxin: A toxic enzyme (protein) excreted into the environment by microorganisms.

extreme halophile: Organisms that thrive in excessively salty environments.

extreme psychrophile: Any distinct group of low-temperature-loving algae, bacteria, fungi, protozoa, and lichens indigenous to the

Arctic or Antarctic. Temperature range is from –18 degrees Celsius to 10 degrees Celsius.

extreme thermophile: Organisms that thrive in excessively hot environments above 55 degrees Celsius.

faculative anaerobe: An organism that can grow with oxygen if it is available, or carbon dioxide if not.

fastidious: Organisms that are extremely difficult to cultivate or isolate due to their special nutritional requirements.

fermentation: Anaerobic oxidation of compounds by the enzyme action of microorganisms. There is no respiration of gases such as oxygen or carbon dioxide required for this microbiological process.

Gas Chromatograph-Mass Spectrometer (GCMS): An instrument on the *Viking Lander* designed to detect, identify, and measure organic molecules and some inorganic molecules such as carbon dioxide and water. It could be used for analyzing surface materials or atmospheric measurements. It is also referred to as the Viking Molecular Analysis Experiment. The instrument's Principal Investigator was Dr. Klaus Biemann, Professor of Chemistry at the Massachusetts Institute of Technology.

Gas Exchange experiment (GEx): One of the three Viking biology experiments used to detect and measure gases coming from microbes in Martian soil samples. The GEx experiment Principal Investigator was Vance I. Oyama of the NASA Ames Research Facility in Mountain View, California.

heterotroph: Microorganisms that cannot use carbon dioxide as the sole source of carbon but require organic compounds instead.

inorganic molecule: A molecule not containing any organic carbon and not usually associated with living organisms.

isomers (L&D): A unifying theme of terrestrial life is its preference for D-carbohydrates and L-amino acids including the L-alanine and D-lactate isomers. When applied to soils on Earth or Mars, the decomposition of one isomer and not the other would indicate living microbes. If both L&D isomers were decomposed, then a chemical explanation in the soil would be the result.

Labeled Release experiment: One of the three Viking biology experiments that looked for evidence of microorganisms in the soil of Mars by placing a drop of radioactive nutrient on the soil and measuring the gas coming out of it. The Labeled Release experiment was designed and built by Dr. Gilbert V. Levin, the Principal Investigator. It was the only experiment on Viking that satisfied all pre-mission criteria for the discovery of life on Mars.

lander: Part of a spacecraft designed to separate from an orbiter in order to land on a planet's surface.

lichen: An alga or sometimes bacteria that live in a symbiotic relationship with a fungus.

lithosphere: The solid rocky outer portion of a planet.

loess: Yellow-colored accumulation of wind-laid particles that are of silt size or smaller.

magnetite: A highly magnetic inorganic mineral found in igneous rocks. It is also internally produced through the enzymatic process of biomineralization in the cells of microorganisms or sometimes outside cell walls as a means of protection from ultraviolet radiation and to help a microorganism orient itself to a magnetic dipole. Magnetite is thought to be abundant in the soil of Mars. It was found along with fossilized bacteria in Martian meteorite ALH 84001 and most recently QUE 94201.

megaregolith: The upper few kilometers of a planet's solid crust that has been pulverized by the bombardment of meteors and asteroids, exhibiting chaotically mixed impact debris. In the case of Mars, this material was a mix of shattered rock and large quantities of ice.

metabolism: The group of life-sustaining processes that includes nutrition and respiration of organic or inorganic material in organisms.

natural selection: A law of evolution first proposed by Charles Darwin in the nineteenth century which says that the greater production of offspring will happen with those organisms best adapted to the ever-changing environment.

Oparin-Haldane hypothesis: The proposal that life arose spontaneously in oceans containing complex organic matter reduced from an atmosphere of hydrogen, methane, and water.

orbiter: Part of a spacecraft that remains in orbit around a planet.

oxidase: An enzyme excreted by a microorganism that brings about oxidation.

oxidation: The process of combining with oxygen. Oxidation can occur as the result of microorganisms secreting the enzyme oxidase into their environment.

ozone layer: A layer of colorless gas created by the dissociation of oxygen molecules absorbing solar UV radiation to produce ozone molecules. The ozone layer then becomes a solar UV shield, blocking the harmful parts of the UV spectrum from reaching Earth's surface. Mars has a very tenuous ozone layer close to the surface.

Panspermia: The hypothesis that life becomes deposited on a planet's surface from lyophilized (freeze-dried) organisms in space that are on the surfaces of meteors and interplanetary dust particles. The theory does not say anything about how life arose.

Patch Rock: A Mars rock image studied by Dr. Gilbert V. Levin at the *Viking 1* landing site with greenish spots on it that seemed to change color with the Martian seasons.

pathogenic organism: A bacteria or virus capable of producing disease.

peroxidase: An enzyme secreted by an organism that catalyzes the reaction of hydrogen peroxide with a reduced substrate, resulting in the formation of water.

photoautotroph: An organism that obtains energy from light (photosynthesis) and uses carbon dioxide as its exclusive source of carbon.

photosynthesis: The use of solar energy by an organism to make high-energy molecules from carbon dioxide, water, and minerals.

phototrophic bacteria: Anaerobic organism capable of photosynthesis, e.g., green or purple sulfur bacteria.

polypeptide: A molecule that contains many joined amino acids.

psychrophile: An organism that is capable of growing at 0 degrees Celsius and below. True psychrophiles cannot grow above 20 degrees Celsius.

Pyrolytic Release experiment (PR): Another one of the three Viking biology experiments designed to detect the synthesis of organic substances in the soil of Mars by microorganisms. The Principal Investigator for the PR was Dr. Norman Horowitz, Professor Emeritus at the California Institute of Technology.

radioisotope: A radioactive isotope, e.g., carbon-14. Sometimes used for the detection of microorganisms in human blood in hospitals and in the case of Mars, for any living organisms. The Viking Labeled Release experiment used a small amount of radioisotope in its nutrient solution, then applied it to a Martian soil sample.

Rayleigh scattering law: A law of physics which predicts that any atmosphere consisting of colorless gases such as oxygen or carbon dioxide will have a blue sky. Most scientists until recently felt Mars only had an orange-red sky and landscape due to suspended dust particles in its atmosphere. When these particles settle out (sometimes for years at a time), the Martian sky is blue or dark purple.

regolith: The loose material above the bedrock on a planet or moon's surface. Regoliths on the Earth are produced by a combination of wind, water, and the activities of life. On the Moon, regolith is produced by the repeated bombardment of large and small meteorites. Regolith on Mars is produced by a combination of the processes similar to the Earth, possibly including life.

saltation: The bouncing of usually sand-sized particles carried by the wind that cause erosion. On Mars, it is not clear that saltation played an important role in the weathering of rocks and the environment because no sand-sized particles were found by the *Viking Landers* at two separate areas 4,000 miles apart on Mars.

snow algae: Photosynthetic microorganisms that are found living in snow and ice, often in the polar regions of the Earth. They can grow at temperatures below 10 degrees Celsius and have a high resistance to solar ultraviolet light. Snow algae colors range from orange, green, red, and blue.

Sol: A term used to denote a Martian solar day or 24.66 hours.

spectroscopy: The science of analyzing the reflected light from a planet's surface, either in the infrared or ultraviolet spectrum. As light passes through a planet's atmosphere (if it has one) certain wavelengths of light are absorbed by the gases present. By comparing the absorption lines and bands in the planet's spectrum with the spectrum of direct sunlight, it is possible to identify the composition of a planet's atmosphere and surface (to some degree).

spontaneous generation: The hypothesis that says life evolves from nonliving matter.

spore: A highly resistant hard-shell seed formed by some microorganisms that can survive large doses of intense radiation and desiccation. Some can even survive in the vacuum of space for indefinite periods of time.

substrate: Any substance used as a nutrient by a microorganism.

sulfur bacteria: Anaerobic microorganisms that use photosynthesis for energy. Growth can be due to either organic carbon or just carbon dioxide, thus rendering the bacteria "faculative," or able to use both. Sulfur bacteria are found to have a large array of colors including brown, green, red, yellow, and purple. They have the unique ability to oxidize sulfide to elemental sulfur and are responsible for large sulfur deposits in the Earth. Mars has 50–100 percent more sulfur in its soil than does Earth.

superoxide dismutase: An enzyme secreted by an organism that catalyzes the dismutation of superoxide radicals to form oxygen and hydrogen peroxide.

terrestrial: A term denoting that which comes from Earth.

triple point of water: The conditions under which ice, liquid water, and water vapor can coexist—atmospheric pressure of 6.1 millibars at temperatures above 0 degrees Celsius. Mars' atmospheric pressure often exceeds the 6.1-millibar restriction (up to 10 mbars) for the triple point of water and therefore can have liquid water on its surface, depending on soil and air temperatures.

ultraviolet rays: Solar radiations from 3900 to about 2000 Angstroms. Ultraviolet radiation in large doses is lethal to macroscopic creatures such as plants and mammals; however, microorganisms have unique adaptations to protect themselves from intense radiation damage such as extracellular biomineral production of iron or magnetite as a shield.

virus: A parasitic microorganism 100 times smaller than a bacterium.

weathering: The interaction and break down of rock by the atmosphere, wind, water, and living things.

X-ray fluorescence spectrometer (XRFS): A *Viking Lander* instrument capable of detecting and measuring inorganic elements such as sulfur, iron, calcium, bromine, potassium, silicon, aluminum, titanium, and others. It was the XRFS that revealed the soil of Mars contained 50–100 percent more sulfur than did Earth soil, and that iron compounds, while abundant on the surface soil, were low or absent under the Martian rocks. It is also referred to as the Viking Inorganic Analysis Experiment.

Index

Page references in *italic* type denote an illustration.

Martin, Leonard, 211, 269
Massachusetts Institute of Technology,
  102–103
mass spectrometry, 130
  see also Gas Chromatograph-Mass
    Spectrometer
Matteia sp., 276
Maunder, E. Walter, 40
MAWD *see* Mars Atmospheric Water
  Detector
Mayeda, Sizuo, 63
measles, 248
media
  and blue sky photograph, 144
  fear of life on Mars and, 306
  fossils in meteorites and, 242, 243,
    249, 264
  NASA representatives and, 174
  reversals of NASA and, 283
megaregolith, defined, 340
Mercury, 23
metabolism, defined, 340
meteors and meteorites
  and comets, relationship of, 35
  contamination and, 177
  fossil microorganisms in, 34, 83,
    202, 241
    announcement of discovery,
      242–243, 249, 250, 264
    controversy of, 334
    discovery of, 256–264, *258, 262–263*
    implications of, 283–285,
      300–301, 317
  lunar organic material and, 131
  Mars origin of, 257, 258–259
  organic molecules in, 83
  Panspermia theory and, 34, 177
  SNC class of, 256–257, 258, 283
  see also comets
Meyer, Michael, 236–238, 283
Michael, Donald N., 250
*Microbial Ecology of Snow and
  Fresh-Water Ice with Emphasis on
  Snow Algae* (Hoham and Duval),
  67–68

microbiology, invention of, 48
microorganisms
  aerobic *see* aerobic organisms
  anaerobic *see* anaerobic organisms
  antifreeze production of, 66, 299
  biomineralization by *see*
    biomineralization
  carbon dioxide and *see under* carbon
    dioxide
  chemoautotrophic, 55–58, 337
  chemolithoautotrophic, 269–270,
    271, 337, *plate 22*
  chemophototrophic *see* sulfur bacteria
  chromogens, 67
  in clay, 191, 192
  color of Mars and *see under* wave of
    darkening
  desert varnish, 70–71, 273, 338
  discovery of, 28
  endolithic, 57, 123–125, 194, 264,
    279–281
  eukaryotes, 171–172
  experimentation for *see* experiments,
    life-detection
  fossils of
    in meteorites *see under* meteors
      and meteorites
    and Panspermia theory, 9
  frozen, 123–125, 223–224
  heterotrophic, 56, 272–273, 339
  lithoautotrophic, 269–270
  lyophilization of *see* lyophilization
  mass of, in Earth's crust, 270
  nanobacteria, 261, 264
  photosynthetic *see* photosynthetic
    organisms
  radiation-resistant, 274
  rocks and *see under* rocks
  scavenging, 120
  size of, and planetary mass, 261
  as sole life form, period of, 171
  structural protection of, 120
  survival studies of, 102–103
  upwelling and outgassing of, 57, 274,
    304